W9-ABS-057

Culture of Bivalve Molluscs

50 years' experience at Conwy

Culture of Bivalve Molluscs
50 years' experience at Conwy

P R Walne

Fishing News Books Ltd
Farnham·Surrey·England

This is a Buckland Foundation
book and is one of a series providing
a permanent record of annual
lectures maintained by a bequest
of the late Frank Buckland.

British Library CIP Data

Walne, P R
    Culture of bivalve molluscs
    1. Oyster-culture—Wales—Conway
    I. Title
    639'.411   SH367.G7

    ISBN 0 85238 063 1

Printed in Great Britain by
The Whitefriars Press Ltd., London and Tonbridge

Contents

List of Figures

List of Plates

7

Foreword

Peter Walne has worked for almost 25 years at Conwy on the development of methods of cultivating oysters and clams. He has been outstandingly successful; the artificial rearing of oysters in commercial hatcheries is now an established fact and there is, at last, a prospect of a re-development of what was once a great industry in Britain. In this book Dr Walne describes the methods followed and the rearing techniques developed and it is possible to distil from this the essence of his success. The Buckland Foundation, which promoted the lectures on which this book is based, commend his work to the close attention of all engaged in the new and promising industry of shellfish cultivation.

Ministry of Agriculture
Fisheries and Food
Lowestoft, Suffolk

H A Cole
Chairman
Buckland Foundation

Symbols and Abbreviations

List of symbols which have been used for brevity in the text:

μm = micron ($^1/_{1000}$ mm)
mm = millimetre ($^1/_{10}$ cm)
cm = centimetre
μl = 1mm^3 = $^1/_{1000}$ ml
ml = millilitre = $^1/_{1000}$ litre
g = gramme = $^1/_{1000}$ kilogramme
kg = kilogramme
ppm = parts per million

Introduction

This book is based on a series of lectures given in London and Dublin in 1968 which described the experiments made at the Fisheries Experiment Station, Conwy, on the culture of oysters and other bivalve molluscs. Although, like other books in this series, it deals with a particular topic relevant to the development of commercial fisheries, it is a little unusual in that it is also very largely a review of the investigations made at one place, Conwy. The interest of the Fisheries Experiment Station in oysters has been a long one, and it is of some interest to trace the origin of this connection. In the early years of the 20th century, the mussels in the estuary of the River Conwy were being seriously polluted and it became necessary, if the fishery was to continue, for them to undergo a purification procedure before they were sent to market. Conwy Corporation took the first step in 1913 by building a set of concrete tanks in which the mussels could be purified, but trials showed that further research was required if the method was to achieve the desired degree of reliability. To this end the tanks were taken over by the Board of Agriculture and Fisheries in 1917, and an extensive series of investigations was begun by the late Dr R W Dodgson. The results of many of these studies may be found in his comprehensive 'Report on mussel purification' (Fishery Investigations, Ser. 2, 10 (1), 1928).

Mussel purification was only required for the winter months, and in 1918 it happened that a few oysters were left in one of the reservoir tanks during the summer. A piece of slate was hanging in the corner of the tank and in early August a few oyster spat were found attached to it. When the tank was drained preparatory to cleansing in readiness for resumption of mussel purification, further oyster spat were found on the concrete walls. From this point investigations into the culture of oysters have been in progress without a break. During the 1920s the work was very largely concentrated on attempts to rear larvae in the outside tanks, and this has been summarized in an extensive manuscript by my former colleague, Mr H P Sherwood, which has been the basis for my descriptions of the earlier work given in

Chapter 2. In the 1930s the work tended towards studies in the outside tanks, supplemented by laboratory investigations designed to obtain more insight into the larval requirements and the basic biology of the oyster. After the war, tank rearing was continued for a number of years, but as the laboratory studies gradually led to a definition of the various conditions required for larval rearing, so the development of intensive culture in hatchery conditions became a reality.

It is fitting that a book on the culture of oysters and other bivalves should appear under the auspices of the Buckland Foundation, since it is a subject that was dear to the heart of Frank Buckland. For many years Buckland was a prolific and popular writer of books and articles on various aspects of natural history. Towards the end of his life (he lived from 1826 to 1880) he was appointed an Inspector of Fisheries, in which capacity he took a great interest in salmon fisheries and in the cultivation of the oyster. His evidence in 1876 to the Select Committee on Oyster Fisheries demonstrates his extensive knowledge based on visits to most of the fisheries, and on his own experiments in pond rearing, which he made on the Kent coast. For further details of the life of this fascinating man the reader is referred to the first series of Buckland lectures given by Garstang in 1929, and to 'The Curious World of Frank Buckland' by G H O Burgess.

The present book is primarily concerned with those facets of the biology of the European flat oyster which are of importance in the culture of their larvae and juvenile stages. Many other aspects are omitted, and for these the reader is referred in particular to 'The Oyster' by C M Yonge (published by Collins in the New Naturalist Series, 1960) which gives a general account of many aspects of oysters throughout the world, and to the outstanding monograph 'The American Oyster, *Crassostrea virginica* Gmelin' by P S Galtsoff (Fishery Bulletin of the US Fish and Wildlife Service, Vol. 64, 1964). The latter, although mainly about the American oyster, also contains considerable comparative information on other species. An earlier book in this series, 'Oyster Biology and Oyster Culture' by J H Orton (Buckland lectures for 1935), also contains a wealth of information.

Detailed references have not been made in the text to all the sources of information. Most of it has appeared in the scientific journals and the relevant articles are listed at the end of each chapter. Where reference has been made to the studies of workers who were not attached to the Conwy laboratory, then acknowledgment is made in the customary manner.

A number of the illustrations are reproduced from other works and the writer's thanks are expressed for these. Permission for their reproduction has been given by: Her Majesty's Stationery Office (Figures 8, 10, 12, 13, 14, 16, 17, 18, 19, 23, 24), the Marine Biological Association of the United Kingdom (Figure 34) and Journal du Conseil (Figure 28).

Fisheries Experiment Station   P R Walne
Benarth Road
Conwy, Caernarvonshire, UK

1 Structure, physiology and reproduction of oysters

In this chapter we are concerned with those aspects of the structure and physiology of bivalves which are particularly relevant to hatchery culture. The descriptions will apply mainly to the European flat oyster (*Ostrea edulis*), with the differences shown by other species pointed out where necessary.

The Bivalvia (colloquially referred to as 'bivalves') are one of the six groups of animals which comprise the phylum Mollusca — an assemblage containing such forms as snails, limpets, cuttlefish, and oysters, clams and mussels. Bivalves are characterized by having a soft body enclosed by two calcareous shells. The power of movement is generally limited and dependent on a single extensible foot; in oysters this is absent and, once past the larval stage, these forms have no power of movement. Within the shells are two ctenidia, generally called gills, which are often relatively large and complicated in structure; they are primarily the food-collecting mechanism.

It is thought that the original bivalves were stocky animals creeping about on a muscular foot, after the fashion of a snail, with a domed shell covering the back. In the course of evolution, the shell became bent down the mid-line of the centre of the back — the dorsal surface in biological terms — and this gradually evolved into two shells hinged on the dorsal surface. After the passage of many generations, the shell became a more massive protective structure, and a life of ready movement was exchanged for one of safety and protection. Much of the sensory system was no longer required, and the head disappeared. Since food could no longer be sought for, it had to be brought to the animal, and this was done by an elaboration of the respiratory gills into organs whose prime purpose is to pump continually large amounts of food-bearing water past the bivalve. There is an abundant fossil record of bivalves, since their shells have been readily preserved in rocks formed from the silty deposits of ancient seas. Bivalves first appeared in rocks of the Ordovician period, and forms of oysters, some of them remarkably like those of the present day, became abundant in the Jurassic and Cretaceous periods, some two hundred million years ago.

The shell of the European oyster comprises two approximately circular valves hinged together on the dorsal side by a horny ligament. The right valve is flat while the left is cupped and is cemented to the substrate when the larva metamorphoses. If the oyster does not become detached, it will spend its life lying on its left side; however, even in natural populations, many individuals do become detached and lie freely on the bottom. There is no evidence that better growth is obtained among those lying either on one valve rather than the other, or in a vertical plane. The *Crassostrea* group of oysters also has dissimilar valves, but the general shape is more elongated in the dorsal-ventral direction. In clams the two valves are similar to each other in shape, and in nature the animal does not lie on one side but is embedded in an upright position in the bottom soil.

Bivalve shells are mainly made of calcium carbonate (over 95 per cent in the oyster) which is laid down in an organic matrix. A section shows that the shells usually contain three layers: a thin outer periostracum which is mainly of organic material, a middle prismatic layer of calcite (a crystalline form of calcium carbonate), and an inner nacreous layer which is composed of a different crystalline form, aragonite. In oysters the periostracum is extremely thin and is soon worn away from most of the shell. The prismatic layer is also reduced and occurs only on the flat right valve, where it forms scales which have a rather horny appearance. Most of the shell of oysters consists of a form of calcite; aragonite only occurs adjacent to the adductor muscle. The shell grows at the border by the addition of material from the edge of the mantle (see below) and in thickness by deposition from the surface of the body.

The general internal appearance and arrangement of the major organs of the oyster are shown in Figure 1. The central adductor muscle, attached firmly to each valve, is a prominent feature and must first be cut before the animal can be seen. The elasticity of the ligament causes the shell valves to gape apart, a characteristic feature of a sick or dead animal. In life the degree of valve opening is controlled by the adductor muscle which is divided into two parts, each having a separate function. On the side nearest to the hinge is the quick muscle which is translucent in appearance; the

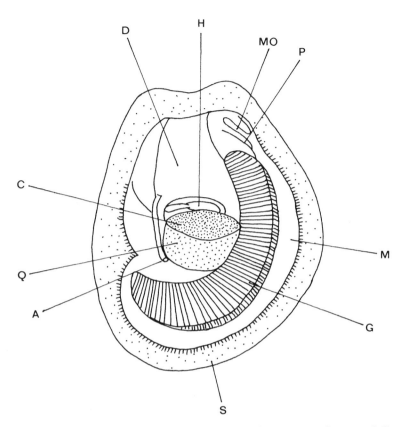

Fig. 1 The appearance of the European flat oyster *Ostrea edulis* showing the principal organs. The flat (right) valve has been removed and the animal is lying in the cupped (left) valve. A, anus; C, catch muscle; D, digestive mass with overlying gonad; G, gills; H, heart; M, mantle; MO, mouth; P, palps; Q, quick muscle; S, shell.

remainder, which is more opaque, is the catch muscle. The former is responsible for rapidly closing the valves when the animal is disturbed, whereas the catch muscle is able to maintain the valves in a closed position for long periods, against the elasticity of the ligament, when conditions are unfavourable. Clams have two separate adductor muscles, and in other species some other variations can be found.

The body mass, lying between the adductor muscle and the hinge, contains the stomach with its digestive diverticula, the intestine, the heart and the kidney. In oysters which are in poor condition or which have recently spawned this area is brown in colour, but generally the appearance is creamy

17

white from the gonad which spreads over the surface. The mantle is a sheet of tissue which hangs on both sides of the body mass and, in life, lies against the shell to which it is fixed at the border. The mantle therefore encloses ventrally a chamber, the mantle cavity. The margin of the mantle is arranged into three folds, two of which are concerned with the secretion of the shell; the inner and largest fold acts as a curtain across the shell opening. The folds on the opposite sides of the mantle, by moving either together or apart, can control the flow of water into the mantle cavity. In those bivalves which burrow, access to the mantle cavity is restricted by the edges of the mantle becoming fused together, leaving only small apertures for the inhalent and exhalent currents, and in forms such as the clam this development becomes extended into two tubes, the siphons. In the course of evolution the original anterior, inhalent aperture has moved to a posterior position so that the two siphons come to lie together. It is the development of elongated siphons which allows many molluscs to burrow deeply into the bottom soil and yet still be able to obtain their food from the water above.

The gills are four crescent-shaped plates which stretch from the mouth for about two-thirds of the distance round the body to a point where the opposite margins of the mantle fuse. This fusion divides the mantle cavity into a large inhalent chamber, containing the gills, and a much smaller exhalent chamber. Oysters of the genus *Crassostrea* have an additional passage for the exhalent current in the promyal chamber. This is an irregularly shaped pocket between the mantle and the right side of the body into which some of the water tubes open. It has been suggested that the presence of a promyal chamber assists *Crassostrea* oysters to live in particularly muddy areas.

The structure of the gill of *Ostrea edulis* is indicated in Figure 2 by means of a section taken at right angles to the surface. This shows how the gill is made up of many minute filaments which are interconnected at intervals and arranged in groups or plica. Although basically each plate is made up of many individual 'V'-shaped filaments there are so many cross-connections that the whole structure forms a firm lattice work. Water is moved from the inhalent chamber of the mantle cavity into the water tubes by the activity of numerous rows of whip-like cilia located on the filaments.

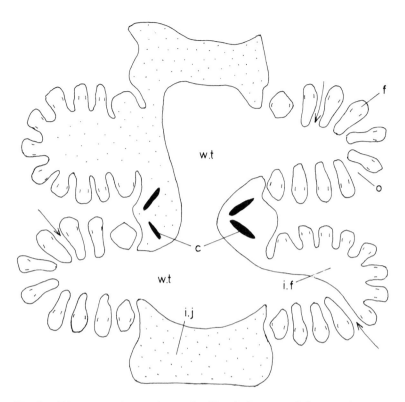

Fig. 2 Diagrammatic section of gill of *Ostrea edulis*. c, chitinous stiffening rods, also shown in the filaments; f, ordinary filaments; i.f, interfilamentar junction; i.j, interlamella junction; o, ostia; w.t, water tube. The arrows show the direction of the water current.

These not only move the water, but also filter from it the small particles which make up the animal's food. The depleted water passes up the gill tubes and so into the exhalent chamber and on out of the mantle cavity.

The structure of the gills of most bivalves is similar, the major differences occurring in the degree of interconnection of the gill filaments. This varies from types such as *Mytilus*, which have no organic connection between adjacent filaments, through that described above for oysters, to that found, for example, in clams, where the degree of connection between adjacent filaments is even greater than that found in oysters.

The gills are not merely sieves; they are also a complex sorting mechanism, and an appreciation of this emphasizes how necessary it is to feed bivalves with particles of the right size and in suitable abundance. As particles are caught by the larger cilia they are thrown on to the frontal surface of the gill filaments. This area bears many mucus-secreting cells and tracts of cilia, some of which beat towards the base, while others beat towards the free margin of the gill. The mucous secretion traps the food particles, and the arrangement of the tracts of cilia is such that the larger particles tend to move towards the free margin while the others move in the opposite direction. On their arrival at either of these points, the particles, and by this time a number will probably have been gathered together by mucus, join ciliated tracts which will carry them towards the mouth. If, however, they are too heavy to be held by the mucous string, they drop off on to the mantle below.

The remaining particles move towards the mouth but they still have to undergo a further sorting process. Each pair of gills ends within a pair of flaps of tissues, the palps, on to which the particles collected by the gills are deposited. The inner faces of the palps (of which there are a pair on each side of the mouth) are folded into numerous ridges and these carry a complicated series of ciliary tracts which separate the particles into small ones, which are carried to the mouth, and large ones, which are carried to the edge of the palp and then drop on to the mantle below.

The rejected particles which fall on to the mantle, either during these sorting processes, or earlier, because they are too heavy to reach the surface of the gills, are moved along ciliary pathways on the mantle to an area close to the inhalent opening. Periodically this rejected material is discharged by a sudden and forceful closure of the shells which is sufficient to blow all the accumulated sediment from the mantle cavity. This rejected material is conveniently termed pseudo-faeces to distinguish it from the true faeces which are formed from the waste material from the alimentary canal. The former are discharged through the *inhalent* opening, against the normal water flow, but the latter are discharged with the exhalent current. Faeces are discharged as short sections of firm ribbon, whereas the pseudo-faeces accumulate as fluffy piles of material with little recognizable form. The production of considerable quantities of pseudo-faeces shows the presence of excessive quantities of suspended material in the water. Such conditions choke the feeding mechanism and if continued for a long period will have a deleterious effect on the animal's condition. In turbid waters, the filtering activity leads to a rapid accumulation of soft mud, which needs to be removed by proper cultivation if the animals are not to become choked and the bottom fouled.

Under favourable conditions bivalves will remain open and pumping for most of the day. The major role of the water current is undoubtedly feeding, but it also carries away waste products from the alimentary canal and the kidneys and brings a supply of oxygen. A pumping oyster removes about 5 per cent of the oxygen content of the water passing through the mantle cavity. During periods of closure, and most bivalves are able to withstand long periods of this, they build up an 'oxygen debt' which has to be repaid when they are able to restart pumping.

The amount of water required by animals of various sizes is clearly of great importance in rearing work, where animals are kept in confinement and have to be provided with adequate volumes of sea water. Estimates of the filtering rate have been obtained by observing the abundance of natural particles in sea water before and after it has flowed past bivalves kept in small plastic boxes. The number of particles in a selected size range can be readily measured electronically with a Coulter counter. The average filtration rates of a wide variety of sizes of five species of bivalves are shown in Table 1. These figures were calculated from experiments made at 20-21°C but, as Figure 3 shows, the filtration rate does not decrease by more than 20-25 per cent at temperatures as low as 15°C. If these data are used to calculate how much running sea water is needed for a batch of animals, it should be remembered that the animals will use only part of the water flowing by. We find that in an efficient system such as the flow trays described in Chapter 5, up to 75 per cent of the particulate matter can be removed by the oysters.

The filtering rate is affected by the temperature, velocity of movement of the water and by the concentration of particles. The effect of temperature on mussels (*Mytilus edulis*) and the Pacific oyster (*Crassostrea gigas*) is similar to that shown for the European oyster in Figure 3, but clams (*Mercenaria* and *Venerupis*) are both more affected by cool conditions. As the flow rate through the boxes was increased so the bivalves increased their filtering rate; a typical example

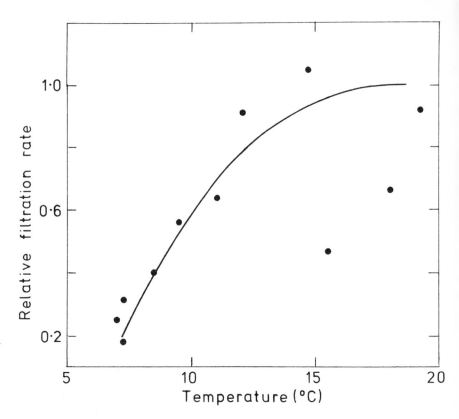

Fig. 3 The relative filtration rate of *Ostrea edulis* at various temperatures, compared with the filtration rate of similar sized oysters at 20-21°C.

is shown in Figure 4. The filtering rate must ultimately reach a maximum value but the method used does not allow this to be measured precisely, since near this point the animal will be removing only a small percentage of the water flowing past it. A series of tests showed that the average filtering rate was increased by about 50 per cent when the flow of water through the box was increased from 50 to 100ml per minute, and by another 50 per cent when it was increased from 100 to 200ml per minute. From this it follows that an adequate flow of water will not only bring more food than a poor flow, but will also stimulate the animal to feed more rapidly.

So far we have considered the mechanism by which bivalves obtain their food and some of the factors which influence this. What is their food? The answer to this

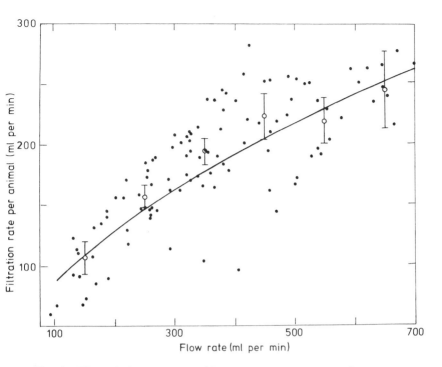

Fig. 4   The relation between filtration rate and water flow rate at 10.8°C for *Crassostrea gigas* with a mean dry meat weight of 0.96 g. The curve has been obtained from the regression calculated after logarithmic transformation of the data. The open circles and vertical lines show the mean value and 95% confidence limits for successive flow groups, 100-199, 200-299 . . . 600-699 ml per minute.

question is that for animals in the sea we do not know. It is clear that the elaborate filtering mechanism is designed to catch very small particles and those down to 2-3 microns (μm, 1000μm = 1mm) in size are removed efficiently; however, smaller particles are less readily abstracted. Very good growth is obtained when both larvae and spat are fed in the laboratory with cultures of unicellular algae (see especially Chapters 4 and 5), but how far this is applicable to young stages in nature is unknown. Although small algae are numerous in the sea, other particles are even more numerous and some of these may be important to bivalve diet.

Once the particles have passed into the mouth, they enter a complicated digestive system which will be only briefly described here. The stomach is an elaborate structure with a variety of grooves and ridges and numerous openings leading

23

to digestive tubules. Mention must be made of one large sac which houses that remarkable structure, the crystalline style. This is a semi-transparent gelatinous rod which lies freely in its sac from which it protrudes across the cavity of the stomach and bears against a hardened area of the stomach wall, the gastric shield. In life, the movement of the cilia causes the style continuously to rotate, one of the very few examples, perhaps the only one, of a rotating part in an animal.

The style liberates digestive enzymes into the stomach and its presence in an oyster is an indication that the animal is in good condition. If an oyster is removed from the water, or is forced to remain closed for some hours, then the style disappears and is not reformed until normal feeding recommences. In other species of bivalves the style sac is more distinctly separated from the alimentary canal, and the style does not disappear so rapidly in unfavourable conditions.

Particles which are not going to be digested pass rapidly through the alimentary canal. In an average sized oyster, material can appear in the faeces within about an hour of being eaten.

Since this book is concerned mainly with rearing the young stages of oysters and other bivalves, the structure of the gonads and reproductive system, and their method of function, must be described in some detail. The gonad in both sexes has a simple form, compared with that found in many animals, and comprises little more than an area of branching tubules covering the outer surface of the digestive gland. The contents of the interconnected tubules are discharged through two small pores opening into the exhalent chamber above the gills. In mussels the tubules are particularly concentrated in the mantle; in clams they spread into the foot. When the European oyster is in a ripe condition the gonad forms a layer 2 or 3mm thick and its creamy colour obscures the brown colour of the digestive gland. In the American oyster the gonad may be up to 6-8mm thick and comprise a third of the total body weight exclusive of the shell.

Not only is the gonad simple in structure, but there are no obvious differences (apart from the presence of eggs or sperm) between the sexes and no doubt this is linked with the ability of a number of bivalves to change sex, a capacity which has been brought to a high degree by the European oyster. When the young oyster first reaches sexual maturity (usually in the summer following settlement, in Britain) the gonad normally develops as a male; after spawning the gonad then changes to the female state, producing eggs instead of sperm, and this regular alternation apparently continues throughout life. The first stages of the production of eggs or of spermatozoa may occur at the same time as the final development of the preceding stage (ie. spermatozoa or eggs respectively). The ripe male stage may develop very rapidly after spawning as a female — within a few days — but no example has been reported of an animal functioning as both sexes at the same time. If this occurred then self-fertilization would theoretically be possible although the oyster may, like many other hermaphrodite animals, be incapable of self-fertilization.

The number of times that an animal will spawn in a season depends on its vigour, and on the water temperature and food supply. In Britain oysters generally spawn twice in a season — once as each sex — but in warmer waters, where the breeding period is extended, several spawnings may be obtained each season. During the winter months no development of the gonad takes place and it may be either completely undifferentiated or else lying dormant in the early male or female phase, ready to resume development in the spring. As we shall see in Chapter 3, this advance can be speeded up by keeping the animals in warm water. In natural waters spawning of *Ostrea edulis* does not usually commence until the water temperature reaches about 16°C. The clam *Mercenaria* and the Pacific oyster require rather higher temperatures.

The regular alternation of sexual development is a striking feature of oysters of the genus *Ostrea*. In *Crassostrea* the sequence of events is different: as in *Ostrea* all individuals first develop as males, but thereafter the population divides into those which remain as predominantly male and those which are predominantly female, although some sex-change can at times occur in later life. Occasional hermaphrodite individuals can be found and examples are known of *C. virginica* and *C. gigas* spawning both eggs and sperm simultaneously. The American clam resembles *Crassostrea* in its sexual development.

In oysters the spawning process and the subsequent

development of the larval stages show some unusual features. It was mentioned previously that both the eggs and sperm are discharged through the genital pores into the exhalent cavity. When the oyster is functioning as a male, the sperm is carried away into the sea by the exhalent current. In the female oyster the eggs pass, not into the exhalent current, but into the water tubes in the gill lamellae; from there they are forced through the ostia, against the direction of flow of the inhalent current, until they lie in the inhalent chamber of the mantle cavity. From this point, in the genus *Crassostrea* the eggs are immediately discharged, by vigorous clapping of the valves, into the sea, where fertilization and subsequent development takes place. The procedure is similar in *Ostrea*, except that the eggs are retained for an extended period, which may reach several weeks in some species, in the inhalent chamber. During this time the eggs are first fertilized by sperm brought in with the feeding current and they subsequently develop into fully shelled larvae before being discharged into the sea, where they undergo the rest of their development. The period when the oyster is brooding its larvae lasts for one to two weeks in the European and the Olympia oyster and is even more prolonged in the New Zealand and Chilean oyster, because the larvae of these species complete their development while lying in the mantle cavity. In this period of brooding the mass of larvae lies loosely on the gill surface. Initially the mass is white in colour and the oyster is referred to as 'white-sick'; this colour gradually darkens during larval development, and the oyster progresses through the 'grey-sick' to the final 'black-sick' stage. The prolonged retention of the larvae is apparently only for protective purposes, since fertilized eggs which have been removed from the mantle cavity will develop normally in glass vessels filled with sea water.

The number of eggs produced by each oyster at a spawning is very substantial. In those forms which shed their eggs directly into the sea the number is much larger than in those which retain them, because the eggs of the latter are larger; the reason for this is that during the period of brooding the larvae have little opportunity to feed, and the egg has therefore to be provided with sufficient food reserves to nourish the embryo through this period.

It is relatively easy to study the fertility of the European oyster, since opening a sample during the breeding season will usually show that 10 to 20 per cent are brooding larvae. The brood can then be collected and the number of larvae estimated. Since there is little or no loss of larvae during the brooding period, counting broods during any stage in development will show the number of larvae which should be liberated. The results of studying a substantial number of samples from oyster grounds in Britain show that the average fertility for successive age-groups (based on growth rates observed at Tal-y-foel) is as follows:

Approximate age (years)	Mean diameter (mm)	Fertility (number of larvae)
1	40	100 000
2	57	540 000
3	70	840 000
4	79	1 100 000
5	84	1 260 000
6	87	1 360 000
7	90	1 500 000

It is probable, although the evidence is not conclusive, that oysters which are in good condition yield more larvae than those whose general condition is poor. The Chilean oyster, which retains its larvae until they are fully developed, produces only about one-tenth of the number of larvae found in European oysters of equivalent size. In contrast, *Crassostrea* oysters produce tremendous numbers of eggs; the number cannot be so readily calculated because not all the eggs in the ovary are shed at any one time, but even so up to 100 million have been recorded.

For a period after fertilization the larva consists of a rounded mass of rapidly dividing cells with little formal structure. In those forms which develop in the sea, the outer surface soon becomes ciliated and the embryo can then move through the water. This is followed by the development of a protective shell of two valves which, within 24 to 48 hours of fertilization, is large enough to cover completely the soft parts of the larva. In the larviparous oysters the rapid early development of the protective shell valves is not so essential, although it is complete by the time the larva is liberated.

The general appearance and the disposition of the organs in the early and in the more developed larva of the European oyster is shown in Figure 5. The most noticeable structure is the velum or swimming organ, a circular lobe of tissue bearing a ring of cilia by means of which the larva both swims and feeds. Small particles caught by the cilia are thrown against the base of the velum and from there are carried to the mouth (the nature of the food is discussed in detail in Chapter 4). The larva alternates between periods of swimming fairly rapidly upwards and slowly sinking through the water column. If the animal is disturbed, the velum can be completely retracted and the two valves of the shell closed. As we shall see in Chapter 3, the closed larva is so very efficiently protected that it can be briefly rinsed with toxic chemicals — such as a dilute solution of chlorine — without being harmed.

The shell length of the larva of the European oyster is usually 170-190μm at the time of liberation, and this increases to 290-300 and exceptionally to 360μm during development. This corresponds to a five- or six-fold increase in weight. It is interesting that the size of fully developed hatchery-reared larvae is substantially greater than is usually found in the sea. It is probable that the reason for this is that the growth rate in the hatchery is often rather greater than in the sea, and maturity is a function not only of size but also of time. Therefore faster growing larvae will have achieved a greater size when they reach maturity. The structure of most species of young bivalve larvae is similar and initially they all have the typical 'D' shape shown in Figure 5. Those which develop in the sea from the egg are smaller: the first shell of *Crassostrea* and of clams measures about 50-80μm.

During the free-swimming stage the dry weight of flat oyster larvae increases about four-fold — from 1μg to 4μg; about 75-80% of this is shell. Some analyses made in 1955 showed that the dry meat of recently liberated larvae contained about 14% glycogen. The variation between broods was slight and was not correlated with the subsequent growth of the larvae. Further progress was hindered because of the difficulties involved in the analysis of the small samples of material which are given even by large numbers of larvae. Recently the NERC Unit of Marine Invertebrate Biology at Menai Bridge has become interested in this problem and their analyses have provided some significant insights into larval development.

Analyses made at Millport and at Menai Bridge have shown that the principal energy reserve of larvae is lipid (oil), not

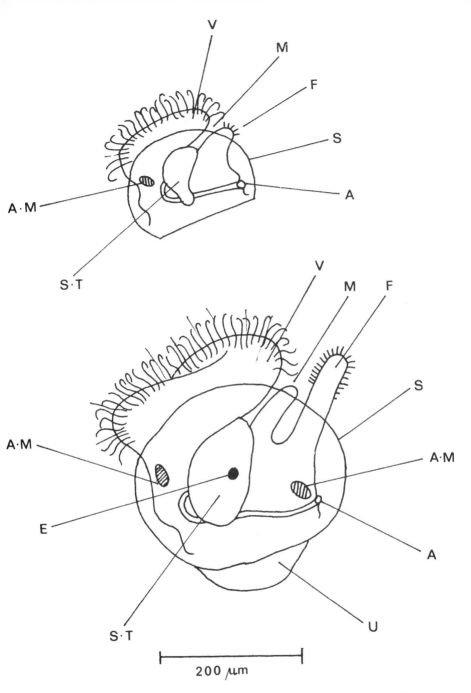

Fig. 5 (A) A newly released oyster larva and (B) a mature larva ready to settle. A, anus; A.M, adductor muscle; E, eye spot; F, foot; M, mouth; S, shell; S.T, stomach and digestive diverticula; U, umbone; V, velum.

glycogen as appears to be the case with adult oysters. Further studies have shown that variations in the lipid content of recently liberated larvae is correlated with their subsequent growth. Samples of larvae obtained over a period of time from two batches of oysters kept in the stock tanks at Conwy had an initial lipid content of 15.5 to 22.5% of the dry meat weight. The corresponding shell length increment in the first four days when the larvae were grown at Conwy in controlled conditions in 2.5-litre beakers increased from 21μm to 55μm. Furthermore, those larvae which grew well in the initial stages also produced a higher proportion of spat. At present these variations in lipid content have come from stock that has been kept over a period of time in our conditioning tanks: oysters which have been kept for a longer period tend to produce poorer larvae with a low lipid content. It is not yet known if there are similar variations in larvae obtained from stocks kept in the conditioning tanks for a similar length of time.

Proximate analyses have also been made of three batches of larvae at intervals throughout their pelagic stage and for a further period as spat. In order to provide sufficient material these larvae were raised in 75-litre bins. These results showed that the larvae accumulated neutral lipid (triglyceride) during development; the proportion increased from 8.8% to 23.2% of the total organic matter. During the period of starvation at metamorphosis there was a sharp drop to about 9.6%. For the first three weeks as spat lipid was accumulated at about the same rate as the other body components so that it remained at about the same proportion — 10%.

As the larva approaches metamorphosis the structure becomes more complex (see Figure 5b) and the shell shape is more characteristic of the species. In the oyster the most noticeable additions are an extensible ciliated foot and a black eyespot (not found in clams) in the middle of each valve. Metamorphosis is a critical stage in the life history, since it is the point in time at which the ability to move is lost, and the internal organs have to be modified to suit a sedentary existence. The behaviour of the larva changes, for if it now encounters a surface while swimming, the velum is retracted and an exploration is made while the larva creeps about on its ciliated foot. After a while the larva may swim off again and this cycle of activity can be repeated many times over a period of a day or two. Eventually a satisfactory position is found (the nature of the surface and other aspects of the behaviour at this stage are discussed in Chapter 4), the larva stops crawling and rocks to and fro on one spot. While doing this it is squeezing out a drop of cement from the byssal gland lying at the base of the foot, and when this has been done the larva applies its left shell valve to the cement. Within a few minutes this sets and the larva is then so firmly attached that a strong jet of water or firm brushing with a soft brush is required to detach it. It has been shown that spat which have settled for only a few hours are able to withstand clean water flowing at 300-400cm per second, but are quickly scoured off if the water contains small sand particles. All species of oysters undergo this attachment process but clams do not; they sink to the bottom and become attached by a byssal thread which can be repeatedly broken and reformed for the first month or so after settlement.

Rapid changes now take place in the tiny oyster. Within 48 hours the velum, foot, eyespot and anterior adductor muscle disappear, the mouth moves through an angle of 90 degrees, and the posterior adductor muscles move to a more central position. The gills first appear as a series of individual filaments which gradually acquire connections between adjacent filaments as the elaborate structure of the adult gill is built up. The mantle becomes more extended and the two valves of the shell grow out parallel to the attachment surface. There is some adhesion between this and the left valve which is eventually lost when the latter starts to form the typical cup shape of the adult shell.

Within 3 or 4 days the basic adult body plan has been acquired and there are no further major upheavals to hinder the oyster's steady development. For commercial purposes it is desirable that the spat should be detached from its settlement surface before it exceeds 2 or 3cm in size, because otherwise the oyster will grow either excessively flattened, or, if the space available is restricted, distorted in some manner.

SELECTED REFERENCES
COLE, H A, 1937. Metamorphosis of the larva of *Ostrea edulis. Nature, Lond.,* **139**, 413-414.

COLE, H A, 1938a. New pallial sense organs in early fixed stages of *Ostrea edulis*. *Nature, Lond.*, **141**, 161.

COLE, H A, 1938b. The fate of the larval organs in the metamorphosis of *Ostrea edulis*. *J. mar. biol. Ass. U.K.*, **22**, 469-484.

COLE, H A, 1941. The fecundity of *Ostrea edulis*. *J. mar. biol. Ass. U.K.*, **25**, 243-260.

COLE, H A, 1942. Primary sex-phases in *Ostrea edulis*. *Q. Jl. microsc. Sci.*, **83**, 317-356.

HELM, M M, D L HOLLAND and R R STEPHENSON, 1973. The effect of supplementary algal feeding of a hatchery breeding stock of *Ostrea edulis* L. on larval vigour. *J. mar. biol. Ass. U.K.*, **53**, 673-684.

HOLLAND, D L and B E SPENCER, 1973. Biochemical changes in fed and starved oysters *Ostrea edulis* L. during larval development, metamorphosis and early spat growth. *J. mar. biol. Ass. U.K.*, **53**, 287-298.

SHELBOURNE, J E, 1957. The 1951 oyster stock in the Rivers Crouch and Roach, Essex. *Fishery Invest., Lond.*, Ser. 2, **21 (2)** 25 pp.

WALNE, P R, 1964. Observations on the fertility of the oyster (*Ostrea edulis*). *J. mar. biol. Ass. U.K.*, **44**, 293—310.

/

2 Tank rearing

Experiments on rearing the larvae of the European flat oyster in the outdoor tanks have taken place at Conwy for upwards of 40 years. During this time many procedures were tested, and a full account would be a book in itself. In this chapter I have had two objects: first, to give an account of the methods and conditions that were in operation when generally successful results were being obtained and, secondly, to indicate some of the pitfalls which were experienced. In the early work, when a run of bad results occurred, there was a natural tendency to ascribe it to the failure of this or that factor. Looking back at the old records, it is often impossible to decide, in the light of present knowledge, what went wrong, since a number of variables, now known to be important, were not recorded. In truth the losses were probably due to combinations of factors, the nature of which varied from year to year, and it was not until a clear idea of the requirements of the larvae had been established that a simple and rational rearing technique could develop.

The site was originally designed for large-scale routine and small-scale experimental purification, and although several kinds of tanks were available for experiment the arrangement was not ideal for testing various rearing techniques, because of the lack of a series of identical tanks. If such a set had been available, it would have been possible to test various factors simultaneously, so giving a greater confidence that the results obtained were related to these rather than to the individual characteristics of the tanks which had to be used.

Description of the tanks

The tanks at Conwy (see Plates 1 and 2) are arranged on three terraces facing the north-east, and their drains are interconnected so that a tank on one level can be filled from any tank at a higher level. When the experiments described in this chapter were made it was only possible to pump directly into those at the highest level, but the pumping main has now

33

been extended so that all tanks can be filled directly from the sea. The original layout consisted of one large tank, measuring 70 × 30 × 6 feet and holding 90 000 gallons, at each of the upper two levels, and two smaller tanks, each 50 × 42 × 3 feet and holding 40 000 gallons, adjoining each other, at the lower level. The larger tanks, known as 'A' and 'B', were those which were principally used for the oyster breeding experiments. At later dates some smaller tanks have been added.

The two large tanks were constructed of reinforced concrete with a smooth cement rendering on the inner surface. Some of the other tanks were built of brick and lined with cement but this was not satisfactory, since inevitably water penetrated between the brick and the cement, causing the latter to come away; most of these tanks have now been replaced with reinforced concrete. The cement rendering gives a smooth, easily-cleaned surface, but it does have to be replaced from time to time; when this is done, the tank requires prolonged soaking before it is suitable for larval rearing. In one instance one of the large tanks was believed to have made the water harmful for larvae up to three months after relining.

The ability to clean the tanks thoroughly at the beginning of the season is regarded as most important. Elsewhere experiments have been made from time to time in ponds, tanks and pits dug in marshes adjacent to an estuary, and used either as excavated or perhaps with the walls supported with bricks, concrete or wood. Experience has shown that these are unsuitable, because the large area of soil sets up unfavourable conditions in the stagnant water. The recent development of large plastic sheets suggests a cheaper construction than concrete tanks; a suitable excavation could be lined with these materials, provided that the soil was sufficiently stable. Cleaning might be difficult but a fresh lining each season might not be prohibitive in cost.

Sea water is pumped from the estuary by a centrifugal pump with a cast-iron body and bronze impeller. The intake is sited at low-water mark on the edge of the main river channel, and the water is delivered to the tanks through 6-inch and 8-inch cast-iron pipes. The tanks are about 2 miles from the mouth of the river and water of suitable salinity can only be obtained for a few hours at high tide. At the time of the original construction it was impossible to avoid metal pipework. This is all of cast-iron (except for the bronze impeller), and from the results obtained it seems to be quite harmless, although the continual addition of rust to the water is a nuisance at times.

It is thought to be an important point that it is possible to obtain water which has undergone very little contamination from intertidal mud flats. Water which has to flow over inshore banks apparently picks up considerable numbers of unicellular, non-motile algae, which are liable to proliferate in the stagnant tanks. These forms are poor food for the larvae and lead to undesirable water conditions. This problem has been particularly marked at Lympstone in Devon, where a number of rearing trials were made in some purification tanks. In this case the intake pipe was a mile from the edge of the channel and non-motile algae frequently developed there.

At Conwy, throughout the period in which these tanks were used for breeding experiments, tank 'A' could be filled directly from the sea, but 'B' could only be filled from 'A'. This difference, and the fact that 'B' is 6 feet lower than and partly shaded by 'A', has had an important influence on the work. As the two tanks are similar in shape, they have been used as a pair with one acting as a control, but it has been observed that in those years when they have been treated similarly considerable differences in the phytoplankton and in the development of the larvae have occurred. These differences have added to the difficulty of interpreting the results of experimental procedures.

Successful spatfalls have been obtained from time to time in most of the tanks, but the most consistently satisfactory results have occurred in the two large tanks, 'A' and 'B', and it is with the experiments in these that the rest of this chapter is concerned. To give the reader a clear picture of the method which was finally adopted, a description of the events in tank 'A' in the summer of 1937 is given in the next section. Essentially the same procedure was used in one or both tanks from 1936 to 1961, and the subsequent sections of this chapter record the various factors that were observed, the difficulties that were encountered, and the ways in which some could be overcome. It should be stressed that this procedure did not suddenly come about but was a synthesis of many trials made in the 1920s and early 1930s.

The experiment in tank 'A' in 1937

At the end of the winter the tank was thoroughly cleaned and the walls and bottom swilled down with a suspension of bleaching powder. This was washed off after a few hours and the tank was then allowed to stand dry for a few days. After this treatment all the attached organisms, which might have flourished during the summer months, had been killed.

The tank was filled on the spring tide period of 5-8 June; after filling the salinity was 32.7 per thousand and the temperature 14.5°C. No additions of water were then made until a partial change in the third week of August, and the tank remained filled until 10 September. A total of 600 stock oysters, comprising 350 obtained from the River Yealm in April, 100 from the same river in June and 150 which had been overwintered in the Menai Straits, were placed in the tank, laid in wooden slatted trays arranged round the sides of the tank and raised about 3 inches above the floor (Plate 4). Each tray contained 30 to 50 oysters, covered with an asbestos sheet; this cover, a few inches above the breeding stock, stopped the growth of filamentous green algae which would otherwise grow rapidly on the shells of the oysters. Before the stock was placed in the breeding tank, the shells were thoroughly scrubbed to remove all traces of weed and sedentary organisms.

Temperature, salinity, pH, the abundance and size of larvae, flagellate density and the daily intensity of spatfall on horizontal plates, were recorded regularly (the techniques employed to observe these conditions are described in later sections of this chapter).

The course of events is shown in Figure 6. A few oyster larvae appeared about the middle of June but substantial numbers did not occur until the period 9-12 July, when the water temperature rose rapidly from 16°C to 20°C. Numbers of larvae remained in excess of 40 per litre almost continuously up to 9 August; it is estimated that in all 218 million were liberated and of these 3.8 per cent reached mature size. During this time the water temperature fluctuated between 16°C and 22°C.

Metamorphosed spat were first found on the test collectors on 19 July, and from then until 8 August spatfalls in excess of 0.5 per cm² in 24 hours were frequently recorded. When the tank was filled, 1500 limed tiles, tied up with string in bundles of ten, had been distributed on the bottom. A

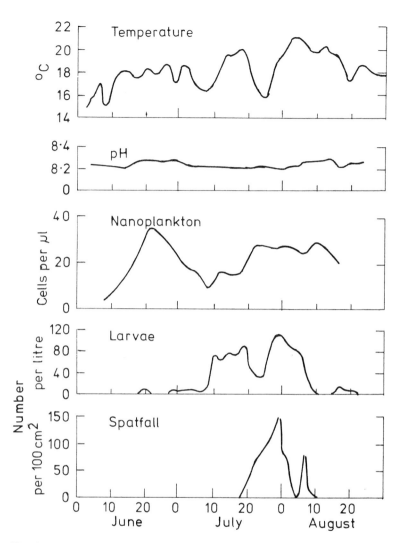

Fig. 6   The course of events in tank A, 1937. (After Cole, 1939.)

further 700 tiles were added in the middle of July, and by
the end of August it was found that all of them (2200) were
well covered with growing spat. Before the end of the season
500 tiles were transferred to a shallow tank where they
received frequent changes of water, but the remaining tiles
were moved directly from the breeding tank to the oyster
ground; later work has shown that this is not so satisfactory
as allowing the spat a period of growth in sheltered tanks

before planting. The tanks were required for mussel purification in the middle of September and by then all the collectors had to be moved to the oyster ground. The nearest suitable area to Conwy is at Tal-y-foel at the south end of the Menai Straits and the work there is described in Chapter 7.

The next logical question, how many oysters can be produced by this procedure, cannot be directly answered. This is partly because of the research nature of the work and partly because of the unsuitable situation of Conwy. In a commercial enterprise, there is no doubt that many more collectors would profitably be used and, of course, they would be kept in sheltered tanks with frequent changes of water until the spat were large enough to be transferred to the sea. In addition the tanks should be adjacent to the planting grounds so that regular attention can be given to the collectors throughout the winter and early spring. From the data obtained it has been estimated that in good years it should be possible to obtain a yield of up to half a million one-year-old spat distributed on 10 000 tiles.

Flagellates

The importance of flagellates as the food of oyster larvae is discussed in Chapter 5; here we are concerned with the estimation of their abundance in the breeding tanks, and the possibilities of exercising control over them.

The species of interest are very small, less than 10μm in diameter, and the most numerically abundant are only 1.5 to 5μm in size. The most nutritious forms are similar in size to human red blood corpuscles. They are therefore difficult to observe and distinguish from the other numerous small particles in the water, although it is easier in the relatively clear water of the tanks than in the turbid waters of many oyster grounds. As the flagellates are not very abundant it is best first to concentrate the sample so as to avoid having to scan large volumes of water under the microscope. It is also necessary to examine freshly collected samples while the flagellates are still alive, since preservatives either alter the cells beyond recognition or destroy them completely. In addition, the characteristic movement of live cells is often important in deciding whether a particle is a flagellate or a detritus particle.

The technique adopted at Conwy for routine flagellate estimation was to concentrate a known volume of sea water by filtration through a collodion membrane, with a pore diameter of 0.6µm; 49ml of the sample was filtered with light suction until the last trace of liquid just disappeared from the surface of the membrane. The organisms left on the surface of the membrane were then resuspended in a further 1ml of sample, thus achieving a concentration of 50 times. Portions of the concentrated sample were then examined on a haemacytometer slide. Doubtless some forms were either left on the membrane or so damaged as to be unrecognizable, but the method has given reproducible results with a number of operators, and, when conditions permitted, with direct counts.

The great majority of organisms seen could not be identified as to species, so recourse was made to a classification based on size and motility. Flagellates were divided into three size-groups: more than 10µm, 2-10µm, and the very small cells of less than 2µm. Diatoms and other non-motile forms (mainly *Chlorella* or similar species) were recorded separately.

It is most useful to consider the abundance of flagellates less than 10µm in size, these being the ones which are suitable for larval food. Larger flagellates were rare; diatoms only occurred occasionally, once the tanks had been filled for a week or two. Other non-motile forms were usually sparse if the tanks had been filled with good quality water. Samples were examined weekly, or often at more frequent intervals, from the two large tanks during June, July and August. In the 18-year period from 1936 to 1953 suitable data are available from a total of 29 tanks; 15 of these were untreated and 14 were treated with organic enrichment (see below). The data for flagellates less than 10µm in size are summarized in Figure 7, so as to show the usual range of concentrations of larval food which were recorded in the semi-natural waters of the Conwy tanks.

In general terms, there is a direct relation between the abundance of flagellates and the spatfall that occurs, although individual exceptions can be found. From 1939 to 1953 the majority of the spatfall occurred in the month of July and in this period there are 26 satisfactory sets of observations of the flagellate population, the pH of the water, and the spatfall recorded daily on a group of 10 oyster shells (spatfall was estimated in a different manner in the earlier years). A useful comparison of these three sets of data

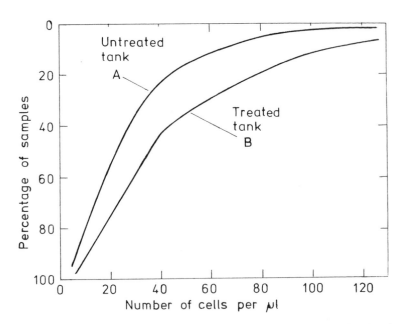

Fig. 7 The percentage of samples which exceeded successive levels of algal cell density in tanks A and B for the years 1939-53 (except 1942 and 1946).

is obtained by plotting the lowest pH (as measured with standard buffer tubes of the indicator Thymol blue and uncorrected for salt error) recorded in July against the maximum flagellate count and the total spatfall for the month (see Figure 8). From this it appears that the chances of a good spatfall are greatly enhanced if the pH is low and the flagellate population is high. On only one occasion was there a substantial spatfall when the minimum pH was more than 8.3 or the maximum flagellate count less than 100 per μl.

The mechanism of the interrelationship of these three factors is not known but the following hypothesis presents a logical argument. Normally the pH gradually rose during the month of July; the lowest figures used in Figure 8 were therefore obtained during the first part of the month. If the pH was high at the beginning of the month, it was usually because the tanks had been filled at the time of a phytoplankton outburst, and it is reasonable to assume that such water would be to some extent impoverished. This would explain the relatively poor flagellate population which such water could support. An abundance of flagellates is

40

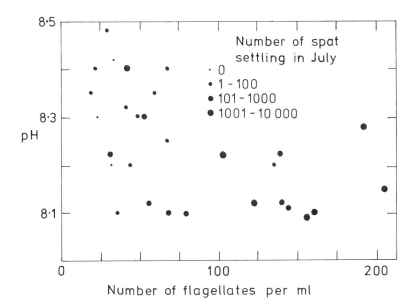

Fig. 8 The relation between the size of the spatfall (represented by the size of the points), the minimum pH and the maximum flagellate count during July, 1939-53. (From Walne, 1956.)

helpful because the population will probably contain a mixture of species, only a proportion of which will be suitable food organisms. Nevertheless, occasions will occur when, although the population is high, good results are not obtained because the useful forms comprise too small a proportion of the species present. Conversely, very good spatfalls may be obtained when the flagellate population is low, if it happens that most of the species present are satisfactory food organisms. Both circumstances have apparently occurred at Conwy, but the fact remains that over the years the spatfall has tended to be greater when flagellates were abundant.

This being so, considerable effort has been expended in seeking ways to enhance the development of substantial flagellate populations. The addition of balanced quantities of nitrate and phosphate always stimulated the production of diatoms rather than flagellates. Much more successful was an organic enrichment in the form of finely ground shore crabs (*Carcinus maenas*). The usual practice was to add a dose of 5-10 crabs a day for periods of up to a week; the total dose was a matter of judgment of the weather and the rate at which the algal population was increasing. Sometimes the algae would increase to very high numbers, which would bring about a high pH and generally unfavourable conditions for larval growth. At other times a considerable enrichment would have little effect; on the other hand, sometimes large populations would develop without any treatment. In general, however (see Figure 7) flagellates tended to be more abundant in the tanks treated with organic enrichment. If a more precise method of obtaining the required dose could be found, this would be a most useful tool in improving the food value of natural water.

At the present time our knowledge of the required nutrient status of natural water for the development and maintenance of a good flagellate population is too limited for this approach. No doubt, with increasing knowledge, this will come and we shall be able to 'cultivate' such tanks in the same way as a meadow can yield a forage crop year after year.

Breeding stock

It is desirable to use stock from several sources, since there seems little doubt that the breeding potential and the vigour of the larvae obtained varies considerably from time to time. At one time some stock was overwintered in the tanks and these invariably gave bad results.

During much of the work 500-600 stock oysters were used in each tank. This was probably excessively high, because good results were obtained in 1961 with only 180 oysters in the tank. The filtering activity of the stock is considerable and must have a substantial effect on the flagellate and larval population. If it is assumed that each stock oyster filters 5 to 10 litres of water per hour, then it can be calculated that a stock of 500 will filter about 25 per cent of the tank volume per day. It is desirable in future work to consider stocking rearing tanks with larvae produced by parents kept separately. The system used in the hatchery work (see Chapter 3) could produce adequate numbers of larvae.

The stock survived well in the tanks; deaths were usually well under 10 per cent. The still water encouraged the development of a substantial shell shoot, but at the end of the season the oysters were in poor condition. Apparently the feeding conditions were not sufficient for the losses due to spawning to be made good. Similarly, it is noticeable that spat do not grow well in these tanks. As soon as collectors are well covered it is desirable to remove them to other tanks where they can have frequent changes of water.

Larvae

The number of breeding stock employed has varied between 200 and 600 and this has always produced a high density of larvae — more than 50 per litre — at intervals during July and August. Although the timing of the first and subsequent major liberations varied according to the weather, the production of larvae has not been a problem. Even in an exceptionally poor summer such as that of 1954 — when the mean water temperature only exceeded $17.5°C$ in 'A' tank for 7 days scattered throughout the summer, and for 12 days in 'B' tank — reasonable populations of larvae occurred, although spatfall was negligible. We have to go back to 1922 to find a year in which the weather was sufficiently bad to reduce larval production. Some batches have appeared to lack vigour — although this is difficult to prove — and to guard against this it has been the practice to stock the tanks from several populations.

Larvae usually measure 170-190μm at liberation. Smaller sizes are sometimes encountered, but these do not usually do well; the larger the larvae, the better they grow. It has been deduced that the duration of the free-swimming period in the Conwy tanks is about 10 days; variations about this figure will occur due to differences in temperature and food supply, and also possibly in the vigour of larvae. The effects of these factors cannot be disentangled because it is not possible to make more than an approximate estimate of the duration of the larval phase. In favourable circumstances a peak can be followed for short periods, but usually the situation is confused by further liberations.

One aspect of larval development which does stand out is the size at which the eyespot develops. This is clearly marked and is presumed to indicate that the major changes which presage metamorphosis are under way. Eyed larvae have been recorded (from the Helford river) as small as 255μm; the largest in the Conwy tanks have measured 350μm. In laboratory experiments in 75 litre bins it was found that eyed larvae varied from 240 to 350μm, and in the development of eyes age as well as size was important. If a population of larvae grew to 280 to 300μm in 10 days only a small proportion would be eyed; if, however, a population took 14 days to achieve this size then, on average, about half would be eyed. It is probable that the relatively large size of eyed larvae in the Conwy tanks compared with the sizes recorded in the sea was due to the improved growing conditions.

Larval liberation and lunar periodicity

Yonge, in his book 'The Oyster', describes the spawning of the female oyster as 'a somewhat surreptitious matter', and indeed it is so. The nature of the stimuli which bring about spawning at a particular time are far from understood, but one that has attracted notice is that of lunar periodicity. Korringa (1957) has found that in the Oosterschelde there is a tendency for larvae to be liberated about 10 days after the new or full moon, and he suggests that the increased tidal range at the time of the new and full moon is responsible. The Conwy tanks are free from tidal influence, but since it has been shown in a number of marine animals that a tidal rhythm will persist when they have been removed from the sea, it was interesting to examine the long series of records of larval numbers to see if any underlying rhythm could be detected. This was done by calculating, from the data of larval numbers and size, the abundance of larvae less than 210μm in length in all the samples examined in the 21-year period 1937-57. The data were than adjusted for each year to a reference date, which was taken as the date of the new moon nearest to 2 July. The result of these calculations, with the curve smoothed by plotting a running average of three days, is shown in Figure 9. From this it can be concluded that any inherent rhythm that may exist in the breeding stock has not been an important influence on the time of liberation of larvae in the Conwy tanks. In practice a rapid rise in water temperature would seem to be the dominating factor.

Spatfall and spat collectors

Earthenware tiles were the usual collectors used in the tank experiments. The original tiles, imported from France, were of the design used on the oyster grounds of southern Brittany; each was about 30cm long by 10cm wide, and curved in a U-shape resembling a portion of guttering. Two holes, about one-third of the distance from each end, allowed the tiles to be tied up in bundles of ten. Sisal twine was used,

44

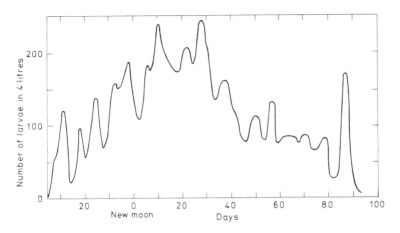

Fig. 9   The abundance of larvae less than 210μm in size in the Conwy breeding tanks for the years 1937-57 related to the new moon nearest to 1 July. Smoother curves drawn from running average in threes.

but as this would not last through to the following summer the bundles had to be re-tied with galvanized iron wire in the autumn, before removal to the oysterage. This could not be used in the summer because of the risk of the zinc corroding and poisoning the larvae. Alternative procedures which were found to be satisfactory included using either soft iron wire or cords of synthetic materials; these could be used in the breeding tanks and would last through the following winter.

The bundles of tiles were dipped in a lime mortar a few weeks before use (Plate 3). The purpose of this was to aid the removal of the spat when they reached thumbnail size, which was generally in the June and July following settlement. The mortar had therefore to be sufficiently strong to remain attached for 9 to 12 months and yet to be friable enough to allow the removal of the spat without excessive damage. A suitable mixture was given by 1½ parts of freshly slaked lime, 2 parts of sand and 1 part of mud. Sufficient water was stirred in to give the consistency of thin oil and bundles of tiles dipped in this were covered with a coating 2-3mm thick. It was important that the coating should not dry too quickly; to aid this the tiles were sprayed with water before dipping, and were shaded from the sun while drying. The mixture should, if possible, be allowed to cure for at least a month.

The coating was very white in colour but a number of tests in which it was stained a variety of colours did not reveal any

45

marked preference on the part of the larvae. The oyster spat were removed during the summer and the tiles were cleaned during the following winter. It was important to remove all the old mortar and traces of marine organisms, otherwise the new mortar would not bond correctly to the tile. For many years this cleaning was done by hand scraping, but a rotating wire brush speeded up the job; the dust problem could be controlled by wetting the tiles first.

The bundles of tiles were well distributed over the floor of the tank. Spat were usually sparse on the underside of the bottom pair of tiles — presumably because of their inaccessibility to the larvae — and on the top of the top pair. The proportion of these poor surfaces was reduced by standing the collectors in banks of two. As the majority were put in the tanks after they had been filled, considerable skill with a long boat hook was required. Banks of three collectors were unstable and piles of three or more could only be built if several were placed together; however, there was no advantage in this, because the lack of water currents meant that larvae could not penetrate into the middle. In several years a wider distribution of collectors in the water column was made by suspending tiles from stainless steel wires strung across the tank. This gave satisfactory results, although the weight of the collectors made the system awkward to handle.

The earthenware tiles had considerable advantages; their shape is both attractive to the larvae and suitable for a number to be stacked together. They are also robust enough to withstand rough weather in sheltered waters, but there are substantial disadvantages; the most important are their weight and brittleness. With lighter collectors, such as those of plastic which were developed for the hatchery work, it would be easier to arrange for them to be distributed throughout the water column but it would be necessary to keep them in protected conditions until the spat were removed. The earthenware collectors were sufficiently robust to withstand winter conditions in the sea.

In Chapter 4 the influence of light on settlement and the attraction of surfaces fouled in various ways is discussed. It is presumed that the natural light regime in the tanks was satisfactory and could not readily be altered, but the degree of fouling of the surfaces offered for settlement could be regulated. The experimental data show that surfaces carrying some natural fouling, including oyster spat, were attractive to settling larvae in the Conwy tanks. This poses a management problem in the following circumstances. Supposing that a light spatfall has occurred, but not sufficient to reach an adequate density on the collectors, and the larval population suggests that a heavy spatfall is expected in the next few days. If further collectors are now added the great majority of the spat will settle on the old collectors. If new collectors are not added, the risk is that the operator will be forced to remove all the old ones in the middle of the spatfall to avoid gross overcrowding. This, and their replacement with new tiles, will cause considerable disturbance of the bottom detritus, and probably a rise in pH, leading to an end of the spatfall. The alternative is to have the collectors suspended in the tank; they can then be changed with much less disturbance compared with those standing on the bottom.

'Trapping' of larvae

This curious phenomenon was liable to occur on surfaces, such as the tiles and the covers to the stock trays, which had been in the breeding tanks for about two months. By then they had acquired a considerable degree of fouling, mostly of unicellular algae and protozoa, and this would stimulate oyster larvae to aggregate on the under surfaces into collections of several hundred to several thousand individuals. The larvae would be in all stages of development and initially apparently healthy, but microscopic examination showed that the aggregations were held together by a slime of bacteria and amoeboid protozoa. The larvae at the bottom of a pile were usually dead, but those at the top appeared normal and active; all sizes were represented.

As each tile could carry several aggregations considerable losses of larvae could occur by this method if not noticed in time. The cause is quite unknown and the only cure was to remove those tiles which were attractive, and to clean and dry them in the sun before returning them to the tank.

Temperature

Although this is a factor of considerable importance, affecting as it does both spawning and larval growth in turn, it is not readily susceptible to control at an exposed site like Conwy. Some experiments were made to simulate the conditions found in the Norwegian oyster polls, where a surface layer of fresh water causes a warming of the underlying salt water. In one trial at Conwy the weather was fine and sunny for 8 days after the establishment of 4 inches of fresh water on the surface of one of the tanks; in this period the water temperature rose to 20.1°C in this tank and in the control. During the following 7 days the weather deteriorated and the temperature in the control tank fell steadily to 16.0°C, whereas in the tank with the freshwater layer the temperature remained at over 20°C. Repeat experiments have given similar results, but the frequency of strong winds usually soon caused the freshwater layer to mix in and be lost. In a more sheltered site a light cover of transparent plastic might be efficacious and would not be too costly.

Sea water and salinity

The tanks were filled on the flood tide at springs, when the water usually had a salinity of about 32 per thousand. It was usual to choose the tides at the end of May or early June so that the water was only stagnant for a short period before the first liberation of larvae. A difficulty encountered in some years was that at that period of the year a bloom of *Phaeocystis* is common in the Conwy estuary. This alga, which takes the form of small irregular gelatinous masses floating in the water, occurs in periods of fine calm weather, when there may be several hundred of these colonies in each litre of water. It was important for filling to be done before this time. The plant quickly died in the tank and formed both a thick foam on the surface, while some sank to the bottom where it decomposed. The algal flora which subsequently developed was generally unsatisfactory. Although the bloom soon disappeared from the river during a period of disturbed weather, the water was apparently impoverished because, if the tanks were filled soon after, the subsequent flagellate population was poor. *Phaeocystis* never grew in the tanks except on one occasion in a 2000 gallon tank; this was continuously aerated from a large diffuser stone, and a substantial population grew and persisted for some time.

Salinity was regularly estimated by determining the spatfall gravity of the tank water with a hydrometer and converting to salinity by the use of Knudsen's hydrographic tables. The values were usually in the range 32-34 per thousand, only exceptionally reaching 35 per thousand.

There is an impression that high salinities are unfavourable but no clear evidence is available; as they are the result of substantial evaporation, other changes may well have occurred. Excessive rise of salinity was controlled from time to time either by the addition of fresh water, or by renewing part of the water with fresh sea water. The latter had the disadvantage that it disturbed the bottom deposit, and a certain amount of dirty water was introduced from the pumping main.

SELECTED REFERENCES

COLE, H A, 1937. Experiments in the breeding of oysters (*Ostrea edulis*) in tanks, with special reference to the food of the larva and spat. *Fishery Invest., Lond.,* Ser. 2 **15**, No. 4, 28pp.

COLE, H A., 1938. A system of oyster culture. *J. Cons. perm. int. Explor. Mer,* **13**, 221-235.

COLE, H A, 1939. Further experiments in the breeding of oysters (*Ostrea edulis*) in tanks. *Fishery Invest., Lond.,* Ser. 2, **16**, No. 4, 51 pp.

COLE, H A and E W KNIGHT JONES, 1949. Quantitative estimation of marine nannoplankton. *Nature, Lond.,* **164**, 694-696.

KORRINGA, P, 1957. Lunar periodicity. *Mem. geol. Soc. Am.,* 67, **1**, 917-934.

WALNE, P R, 1956. Observations on the oyster (*Ostrea edulis*) breeding experiments at Conwy, 1939-1953. *Rapp. P.-v. Réun. Cons. perm. int. Explor. Mer.,* **140 (Pt. III)**, 10-13.

3 Hatchery rearing of oyster larvae

In Chapter 2 an account was given of the methods used to rear oyster larvae in the outdoor tanks during the summer months. Here we are concerned with the production of oyster spat in relatively small tanks indoors. Since these are both small and indoors it has been possible to envisage a more rigorous control of the conditions than was possible in the semi-natural environment outside, but, before limits could be set to these, very detailed studies have been necessary on all the factors of potential importance.

At the same time as the tank rearing experiments were being made laboratory studies were also in progress. The work was concentrated particularly on the food requirements and the preliminary studies at Conwy were expanded into a pioneer study at the Marine Biological Station at Port Erin. This work, reported by Bruce, Knight and Parke in 1940, was an important milestone in the progress of oyster culture. Here was a clear demonstration of the importance of the nature of the food and the elaboration of a technique of culture.

After the war the work was extended at Conwy to cover other factors, and by the beginning of the 1960s the methods for rearing the larvae in the laboratory were sufficiently reliable for them to be expanded to a larger scale. The techniques then became not only a research tool but also an alternative method of producing substantial numbers of young oysters for commercial purposes.

The enhanced scale of work required a long period of technological study. Operations which are simple when the populations of larvae can be measured in hundreds, and the vessels contain only a litre or so of water, become a substantial task when the vessels increase to 50 to 400 litres and the larvae to hundreds of thousands. Equipment which will withstand the corrosive action of sea water and yet be harmless to larvae cannot be readily found.

For these reasons this chapter not only describes the handling of the larvae but also the important ancillary services: the seawater supply, the provision of adequate amounts of algae for food, and the care of the breeding stock. The breeding work is now carried out in a Culture Unit which was specifically designed for the purpose and is sited alongside the tanks described in the previous chapter; it contains, as well as other laboratories, a tank room for the maintenance of the breeding stock, an algal culture room, and a controlled temperature room for the culture of the larvae. This separation is desirable for research purposes and also for the larger scale of work in a commercial unit. Chapter 4 discusses in more detail a number of the factors which influence the well-being of the larvae.

The seawater system

The provision of good quality sea water is most important, and since the regular cleaning of all the tanks and pipes in the system is essential for good growth and survival of larvae, the design should allow this to be done easily. It is helpful to keep all pipe runs as short as possible.

A 20 000-gallon reinforced tank is used as the main reservoir and it takes about half an hour to fill at high tide. In stormy weather it is better to pump first into an adjacent 90 000-gallon tank and to transfer the water after a day or two when some of the silt has settled. The storage tank is cleaned out every fortnight, when the walls and floor are scrubbed and hosed; in addition they are swilled down with bleaching powder solution once a fortnight in summer and once a month in winter, to control the growth of weed.

Water is pumped, as required into the laboratory block through a 2-inch polythene pipe into a set of three interconnected 100-gallon fibreglass tanks in a loft on the roof. The pipe is made up of short lengths which are connected with screw fittings so that it can be readily taken apart for cleaning. More recently we have found that the pipe can be cleaned with an abrasive plug forced through by compressed air. This is easier and more economical in time, but some modification to the pipe so as to eliminate sharp bends and other obstructions has been necessary. The pump, which has a stainless steel rotor and a rubber stator, is sited below the level of the water in the storage tank; this helps to avoid priming trouble. As there is no foot valve on the suction side, the pipe empties to the level of the water in the tank when the pump stops. Although this helps to prevent the growth of sedentary organisms, a bacterial slime soon appears and the pipes are cleansed once a fortnight in summer.

Since rearing work is in progress throughout the year, substantial quantities of sea water have to be warmed. For several years this was accomplished in a fibreglass tank adjacent to the loft storage tanks. This was fitted with three 3-kW immersion heaters sheathed with 'Vitreosil' — a silica material that is inert in sea water; they were controlled by a thermostat inserted into a glass thermometer pocket fixed vertically in the tank. This system delivered about 200 litres per hour heated to 22°C from an outside temperature of 2-4°C. Electricity is, however, an expensive fuel and this has now been supplanted by a gas heating system. The sea water is heated in glass heat exchangers fed from the primary circuit of a gas boiler. The temperature is regulated by motorized valves which turn the primary circuit on or off according to the demand of a thermostat fixed in the seawater supply. Warmed sea water is supplied from these to the tank room and the hatchery in PVC pipes.

In the hatchery, the sea water is further treated before it is suitable for use in larval rearing. Repeated trials have shown that better growth is obtained if suspended silt and competitive organisms are removed by filtration. This is illustrated by these two separate experiments in which larvae were grown for 4 days in glass beakers. One set of two contained 'natural' water from which the larger particles had been removed by a plastic sieve with a pore size of 63μm, while in the other set the water had been filtered by the procedure described below using a sand filter and ceramic candles. *Isochrysis* was added to all vessels as food for the larvae.

Initial size of the larvae (μm)	Average increase in shell length in 4 days (μm)	
	Natural water	Filtered water
206	46	63
228	35	42

In the first small-scale experiments ceramic filter candles were used. As the experiments grew larger the rate of filtration was increased by pumping the water through tubular ceramic elements enclosed in ebonite or PVC holders. We used elements with a nominal pore size of 25-30μm. These gave good results but they speedily became choked if the water contained much silt or plankton, and it took some time to unbolt the holder and fit a clean element. We found that their life was improved if the water was first passed through a simple sand filter, and water treated in this manner was satisfactory for larval culture.

By this stage the apparatus had become rather elaborate, and a good many plastic pipes and cocks were needed for its operation. All these were potential foci for the growth of bacteria, and unless there was a high degree of maintenance the filtered water contained more bacteria than when it entered the system. This was improved by pumping water containing 3 ppm of chlorine, obtained from a commercial hypochlorite solution, through the system at the end of the day's operation, but even this, as well as other precautions such as frequent changing of the sand and ceramic filters, did not result in water with an exceptionally high degree of bacterial purity.

Water filtration has now been considerably simplified by the installation of a plate filter which uses Kieselguhr — a powder prepared from the siliceous shells of fossil diatoms — as the filter medium. All parts of the filter which come into contact with the sea water are made from a plastic material. The apparatus can be readily taken down and thoroughly cleaned each day, and by varying the number of plates (each 30cm in diameter) the capacity of the filter can be altered according to the volume of work in hand (Plate 8).

Some data of the filtering efficiency have been obtained by counting the number of particles in the water with a Coulter counter before and after filtration. With this electronic instrument it is possible to obtain an estimate of both the number of particles and their size. About 50 samples were analysed at intervals in 1966 and the results, set out in Figure 10, show that the original sea water usually contain 10 000 to 20 000 particles larger than 3.2μm in diameter per ml, and this was usually reduced to about 2000 particles by the filtration procedure.

After the water has been filtered it is passed through an ultra-violet sterilizer; this consists of a 44 W low-pressure mercury discharge tube in a quartz envelope. Passage of the filtered water in the annular space between the envelope and

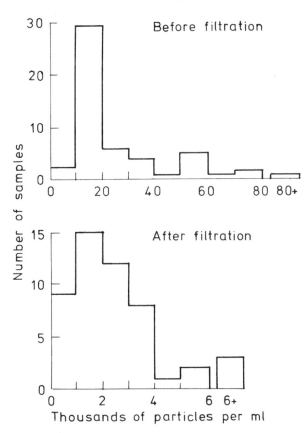

Fig. 10   The number of particles per ml larger than 3.2μm in diameter in 50 samples of sea water used in the oyster hatchery before and after filtration. (From Walne, 1970.)

an outer glass jacket at the rate of 300 litres per hour can kill more than 99 per cent of the bacteria, but this is in an ideal situation where a very high degree of filtration and pipe cleanliness has been obtained. In practice the results are less good. The following table shows the average results from weekly estimates of bacterial numbers, obtained by plating samples of water with ZoBell's agar media no. 2216, over a 9-month period. The counts of the water in the reservoir tanks are taken as 1 (they varied between 16 and 10^6 per ml) and all the other observations are expressed in proportion:

Untreated sea water	1
After filtration	0.36
Delivered to the larval culture bins	0.07
(after passage through the ultra-violet unit)	

The piping between the filter and the sterilizer, and from there on, is either glass or heat-sterilizable PVC. These are cleaned daily either by the passage of hypochlorite in fresh water or by steaming.

The result is clear; only by careful working and a substantial amount of care is the bacterial content of the water reduced to about 5 per cent of that in the original raw sea water. Of course the composition of the flora may be different due to both selective removal and re-infection. The readiness with which sea water can become infected with very large bacterial populations is not appreciated by many people, and the following example illustrates this point. An uncovered beaker of autoclaved — and hence bacteria-free — sea water was placed on a bench in the hatchery. Twenty-four hours later, when a cover was put on, it contained 1800 bacteria per ml and these had increased to 500 000 in 48 hours. The original contamination must have come from dust and water droplets in the air.

Algal cultures
Large quantities of uni-algal culture are required to feed the breeding stock, the larvae and the spat. At first a rather simple system was used in which algae were cultured in glass aquaria with lights shining through the sides. Although substantial yields of vigorous species such as *Dunaliella* and *Phaeodactylum* could be obtained, the more delicate forms — which are the preferred food of the larvae — soon succumbed to contamination. Since reliable supplies of culture are essential for hatchery work a more controlled system was developed. This combined a fairly high degree of reliability with a minimum of labour. Since, for research purposes, we also need to grow a number of species at the same time, an apparatus using 14 culture vessels was built. In a commercial hatchery their management would be simplified since only a few, but larger, vessels would be required.

The algae are grown in glass flasks holding 20 litres of sea water enriched with nitrate, phosphate, other inorganic compounds in trace amounts (see Table 2) and 2 vitamins (Plates 6 and 7). The tops of the flasks have a flared neck and this, in conjunction with a backing flange, allows a stainless steel plate to be bolted over the mouth. Four tubes pass through this plate: one for the addition of fresh medium, an aeration tube, a harvesting tube, and a vent to allow air to escape. These are grouped round a central bearing through which a stirrer shaft is passed.

The cultures are brightly lit with fluorescent lights, the heat from which is dissipated by standing the flasks in a trough through which cold water flows. A mixture of air and carbon dioxide, which is sterilized by filtration through cotton wool, bubbles through each culture at the rate of 50 litres per hour to maintain a pH of 7.5-8.0. If the pH rises to a high level growth will be reduced.

The use of natural sea water in the medium involves some risk since the water will contain many contaminating organisms which have to be removed by a sterilization process. The alternative is to use artificial sea water as a basis and this has some attraction because the quality will not vary (as ordinary sea water does) and the contaminating organisms which have to be removed by the sterilization process will be much fewer, but its preparation in large quantities presents some practical problems since sea water contains about 3 per cent of salts. A quantity of 100 litres will contain a total of 3kg of salts which have to be brought into solution.

The usual seawater medium is prepared in batches of 100 litres, and after sterilization it is kept in a glass flask from which it is drawn as required. At one time we used heat for sterilization. The flask was fitted with an electric immersion heater coated with 'Vitreosil', a thermostat and a stirrer. The batch was heated to 90°C (taking about 9 hours) and held at this temperature for a further hour. Although this method was reasonably effective, the water took a long time to cool down, during which time a considerable precipitation of the more insoluble salts of sea water took place. This method has now been replaced by a filtration system; after a preliminary coarse filtration sea water is pumped from the laboratory supply to a tank on the roof where the necessary enrichment is added. From this tank the water flows by gravity through a glass filter press fitted with a sterilizing grade of a cellulose-mineral mat, into a large reservoir flask. The press had eight pads, 14cm in diameter, and with the hydrostatic head of about 36 inches a batch of water takes about 20 hours to pass through. Larger presses or models with more pads are obtainable if more rapid filtration is required.

The cultures are harvested by removing, through a siphon, 5 litres at a time and this volume is then restored by the addition of fresh media. Each vessel can be harvested at this

Plate 2   The outside tanks. The large tanks at the centre and right were used for the outdoor culture experiments. The shallow tanks on the left are used for mussel purification.

Plate 3   Coating the earthenware tiles with a mixture of lime and sand.

Plate 4   One of the large culture tanks before filling. The breeding stock are in the trays arranged against the wall of the tanks.

Plate 5   *Ostrea edulis* breeding stock held in running heated sea water.

Plate 6   Algal cultures maintained in Erd-schreiber medium.

Plate 7   Twenty-litre continuous algal cultures.

Plate 8   The equipment used for preparing the water for larval culture. The filter press is in the left fore-ground with the glass heat exchanger behind and the ultra-violet steriliser above.

rate about two days out of three, and to keep the culture in good condition it is essential that this is done, whether or not the food is required. Over a period of 100 consecutive weeks the yield of *Isochrysis* was 29.2 litres per week of culture containing 10 000 cells per μl (a cubic millimetre). This yield is an average figure drawn from a number of cultures, for during this time individual cultures lasted for about 52 days. Despite the attention given to sterilizing the media and regular cleaning of all the pipe lines by blowing steam through them, infection remains a serious problem. The main contaminants are ciliates and colourless flagellate, *Monas* sp., both of which eat algal cells. Small colourless flagellates also appear from time to time, and these have been shown to come through with the medium. *Tetraselmis*, which is also used for feeding the larvae, is a larger species of flagellate and is more resistant to contamination; cultures of it generally last for upwards of 100 days and yield about 30 litres per week at a cell density of 1000 cells per μl. In terms of algal volume this is similar to *Isochrysis*.

In practice we find that it is better to regularly change the cultures on a rotational basis, to help maintain them in a vigorous condition. Despite this precaution difficulties still remain in maintaining vigorous cultures, particularly of *Isochrysis*, at all times. There are periods when apparently the sea water itself is unsuitable for supporting a vigorous algal growth. At those times, instead of *Isochrysis* reaching more than 10 000 cells per μl, a maximum cell density of only five to seven thousand is the highest attained, even with a reduced rate of harvesting. It is possible that the reduced yield is caused by changes in the bacterial flora rather than in the chemistry of the water. If so, the difficulties could be overcome by rearing bacteria-free cultures, although in practice this presents a formidable task as well as a considerable additional capital outlay. Some trials made several years ago at Conwy did not produce better growth when the larvae were fed on bacteria-free cultures. The use of artificial sea water is probably more readily applicable at those times when the results with natural water are poor. *Tetraselmis* grows very well in artificial sea water made up to the formula given in Table 3, but the growth of *Isochrysis* is less good.

Cultured algae are required for three purposes in a hatchery: feeding the breeding stock, the larvae and the spat.

There is no doubt that the larvae are particularly sensitive to the quality of their food supply, and it is very worthwhile to ensure that this is available as required. It is, however, probably the most technically difficult part of hatchery rearing, so it is expedient to reduce this part of the operation as far as possible. This can be done by separating the culture of the larval food from that required for the other stages, since these are better able to withstand variations in quality and quantity. Methods are available for growing some species of algae in vats or plastic bags; these work well, provided that a dense initial inoculum is used and the culture is not kept going too long, since contaminants are able to get in more readily than in more controlled conditions.

The use of fibreglass vats holding 300 litres of sea water was introduced by the W.F.A. and we have found this a useful advance towards the large-scale production of *Tetraselmis* culture. These large vessels do not give satisfactory results if used on a semi-continuous basis, but good results are given with a batch culture system. Ten litres harvested from the 20 litre cultures are used for the inoculum and the culture usually reaches 1000 cells per μl in 6 to 7 days. Fifty litres a day is harvested for 3 days and 150 litres for the last day. In 25 consecutive runs, 12 per cent partially failed due to contamination which is always a problem in the less closely controlled systems. In the remaining 88 per cent the average yield from each vat was 412 litres at 1000 cells per μl and each culture lasted on average for 10.6 days.

Breeding stock

During the summer months oyster larvae can be readily obtained by keeping a dozen oysters in a 10 gallon tank and either changing the water frequently, or using a continuous flow of fresh sea water (Plate 5). Any larvae that were being brooded will be released in a few days. If, however, one wishes to rear larvae at those times of the year when brooding females cannot be obtained from the sea, then more elaborate arrangements have to be made. To obtain larvae in the winter months the stock has to be kept while the gonads ripen, through spawning and during the subsequent development of the larvae. Although rigorous proof is lacking, it is logical to provide good conditions for the stock during this stage, to help ensure the development of larvae of good quality.

It is desirable to use oysters from a variety of places since there are marked differences in the readiness with which different populations spawn, and in the subsequent vigour of the larvae. However, if this is done precautions have to be taken to ensure that pests, not found at Conwy, do not escape with the effluent water. The Molluscan Shellfish (Control of Deposit) Order 1965 prohibits the movement of oysters from certain areas where pests occur to other specified areas except under licence. The tank room at Conwy has two bays, each draining separately to the outside. The bays, separated from the centre of the room by concrete curbs, contain single rows of aquaria arranged in two tiers. The method adopted to isolate the breeding stock has been to arrange so that one of the bays drains into a 2000 gallon concrete tank outside the building where the effluent is collected and chlorinated before discharge to the sea. The procedure when a tank is full is to divert the flow to another tank, and then stir in sufficient of a slurry of bleaching powder to give a concentration of 100 ppm of chlorine. This is then allowed to stand overnight before discharge.

At first iron-frame glass aquaria were used, but these have now all been replaced with polythene boxes which are cheap and long lasting. Sea water, warmed to 22°C, is distributed to the aquaria from a manifold made up of a series of screwed polythene fittings. Because warm water is continuously flowing through these pipes, various sedentary organisms, which come in with the sea water as larvae, settle and grow. The use of screwed fittings allows the system to be quickly dismantled and cleaned.

The supply of water to each aquaria is controlled by a screw clip and runs directly on to the surface at one end of the tank. The parent oysters are held at about mid-depth on a plastic grid hanging from cross-bars resting on the top. A circulation of water is ensured by having the overflow through a tube reaching to the bottom of the tank, at the end opposite the inflow. This has the additional advantage that, because newly-released larvae swim strongly to the surface, they tend to be trapped in the tank.

Each tank is usually stocked with 10-15 oysters, and a flow rate of 10 to 15 litres per hour is maintained. This is certainly less than that utilized by animals in the sea but to increase it would raise the heating costs and would also increase the risk of washing the larvae away. A screen on the overflow has been tried but the results are not commensurate with the work required to keep it from becoming choked. Another way in which many larvae doubtless are lost is through being eaten by the filtering activity of the parent stock.

Because of the low flow rate, and to guard against periods of the year when the natural flora is reduced, the warmed sea water is enriched with algal cultures. For some time this was achieved by automatically dosing small quantities of culture into a header-tank into which the warmed sea water was delivered before distribution to the aquaria. This has now been superseded by a much neater arrangement where the algal culture is injected directly into the seawater supply pipe. A peristaltic pump is used and its rate of dosing is adjusted so as to empty a 20 litre reservoir in 24 hours. This reservoir is filled daily with any available culture. At one time two vigorous species *Phaeodactylum* and *Dunaliella* were specially grown for this purpose, but as we now know that these are not ideal foods *Isochrysis* and *Tetraselmis* are mainly used.

When the larvae are liberated they are readily seen, because they rise to the surface where they congregate to form grey flecks. They can then be collected by sieving the water through a fine-mesh sieve. So far no method has been found which will stimulate spawning, although there is some indication that it can be stimulated by a rapid change of temperature. It has been noticeable that it is common for liberations to occur in the afternoon. In 1962-64, 173 liberations of larvae were recorded and 100 of these were found in the afternoon compared with 73 either overnight or in the morning.

Rearing the larvae
As the investigations proceeded it became clear that scrupulous cleanliness was necessary for satisfactory rearing, and to aid this all the large-scale rearing and many of the smaller experiments are carried out in a room set aside for the purpose — the hatchery. This room has no windows and the air temperature is controlled so that all the vessels can be placed where convenient. Generally larvae are reared at 20-22°C. A gas-fired warm-air heater placed outside the room has proved very convenient. Heating equipment in the hatchery itself is a nuisance because it is frequently splashed

with water and is difficult to clean. Much of the earlier work was done in a glasshouse and it was found that in these conditions cultured food can 'bloom' when added to the larval cultures, and undesirable pH conditions can result. Light does not seem to be necessary for larval development and it seems best to exclude it and so reduce the number of variable factors.

Larvae have been reared in a great variety of vessels, varying from those holding several hundred litres of water to beakers holding one litre or less. The greatest experience with large-scale rearing has been obtained with polythene bins holding 75 litres (Plate 9). The use of these vessels was examined in detail in the years 1963-64 and again in 1969, and the following paragraphs outline the procedure adopted and the results obtained. The process is still in general use for rearing large batches of larvae.

The bins are 23 inches high with a diameter of 15½ inches at the base and 18 inches at the top. They are filled with 75 litres of filtered sea water and sufficient antibiotics added to give a concentration of 50 international units of penicillin G and 0.05mg of streptomycin sulphate per ml. The yellow-brown flagellate *Isochrysis galbana* was used exclusively as food, and enough culture, usually 500-750ml was added to give a cell density of 100 per μl. Further study has shown that a mixture of 50 cells of *Isochrysis* and 5 cells of *Tetraselmis suecica* per μl gives better results (see Chapter 4) and this was the food provided throughout 1969.

One hundred thousand freshly liberated larvae are added to a bin, the contents of which are gently stirred with a stream of air bubbles from five aerators hanging near the bottom of the bin; the aeration is standardized at 200 litres per hour (Plate 10). The water is completely changed on alternate days. This is done by siphoning the water through a filter formed by a bag of nylon bolting cloth with a mesh size of 124μm which retains the larvae (Plate 11). When the water has been reduced to an inch or so in the bottom of the bin, the filter is washed to remove any adherent larvae and the contents of the bin are tipped into a bowl. Although larvae are sensitive in their requirements, they are physically surprisingly rugged and the next step in changing to fresh water is to concentrate them on to a sieve (Plate 12). At first stainless steel sieves were used, but these are expensive and do not usually, because of corrosion, last more than a season.

Now we make our own by gluing nylon bolting cloth on to sections of PVC pipe with PVC cement.

The larvae are first washed through a sieve of 295μm mesh which retains the large pieces of debris and then a jet of water is directed on to them as they lie on a sieve of 124μm; this washes the smaller particles away. At first, this procedure was altered on the fourth day when, in addition, the larvae, after washing, were stood in a beaker containing a litre of sea water to which sufficient of a commercial hypochlorite solution was added to give a concentration of 3 ppm of chlorine. The larvae were left for five minutes and this killed most of the bacteria adherent to the shell. The purpose of this step was to try and prevent the larvae carrying over on their shells inoculations of bacterial populations which had been building up in the culture and could include pathogenic forms. Since brief chlorination does not apparently affect the larvae this step has since been superseded by rinsing with a chlorine solution at every water change.

When the larvae have been washed they are replaced in a clean bin containing freshly filtered water, antibiotics and food. On those days when the water is not changed, the food content of the water is checked by examination of a sample under the microscope and sufficient algal culture is then added to restore the food content to its former value. Usually about one-third to one-half of the food has been eaten overnight.

When 'eyed' larvae (see Chapter 1) are seen two strings of five mussel shells are hung near the surface. A daily index of spatfall is thus obtained by counting the settlement on these shells which are then cleaned before being replaced. When the test shells show that spatfall is beginning, the main collectors are added. These are formed from sheets of black matt-surfaced PVC 0.030 inch thick which have been moulded to the same size and shape as the earthenware tiles used on the French oyster grounds. They are tied together in bundles of 8 or 16 with polythene twine and stand on the bottom of the bin; a bundle of 16 reaches nearly to the surface.

Spatfall continues for about 7 days. After this the daily spatfall is very small, although about 30 per cent of the original larvae may still be present. The collectors are then removed; in very productive batches it may be necessary for them to be changed before spatfall is complete. It has been found that substantial mortalities take place if the spat are placed directly in natural sea water, and so for some time it was the practice to place them in filtered sea water enriched with the same food as the larvae; antibiotics were omitted at this stage. More recent work has demonstrated that better results are obtained if the spat are kept in coarsely filtered water enriched with algal culture. The most satisfactory method has been to stand the collectors in fibreglass tanks holding about 400 litres and filled with sea water filtered through a 68μm mesh; good growth is obtained if this is enriched with either 50 *Isochrysis* or 2.5 *Tetraselmis* cells per μl. Such a tank holds 6 to 8 bundles of collectors, and to help promote a circulation of water we use two tanks, one above the other, and the water runs slowly from the top to the bottom tank. A pump, controlled by a float switch, returns the water to the top. In addition it is often necessary to add algal culture continuously because of the intensive grazing by the spat. When these have grown to 5-10mm the collectors can be put outside and their fate at this stage is dealt with in Chapter 5.

In 1969 the procedure for collecting the spat was modified in two ways: a disc of PVC lying on the bottom of the bin was the only collector used, and the metamorphosed larvae were brushed from this each day (Plate 13). Comparative trials have shown that as many spat could be gathered from a collector arranged in this manner as on a stack of 'tiles', and provided that they were brushed off within 24 hours of settlement very little damage was caused. If the number of spat are not too many they can be brushed off with little damage but if there is a heavy spatfall many of the shells are broken during removal and the spat die. This difficulty can be overcome by using the blade of a sharp knife to cut between the shell and the plastic. These very small spat, usually described as culchless spat, were then kept in trays with a mesh base through which filtered sea water, enriched with algal good, was pumped continuously. This newer technique, and the results are discussed in more detail in Chapter 5.

Results of hatchery rearing
The results of a series of experiments made in 75 litre bins during 1963-64 are outlined in a series of diagrams (Figures 11 to 14). These show respectively the average growth rate of the larvae, the time of development of the eyespot

(indicating the presence of mature larvae), the food consumption of the larvae, and the course of spatfall. In general the greatest number of mature larvae were present 14 to 16 days after liberation, when spatfall also reached its peak.

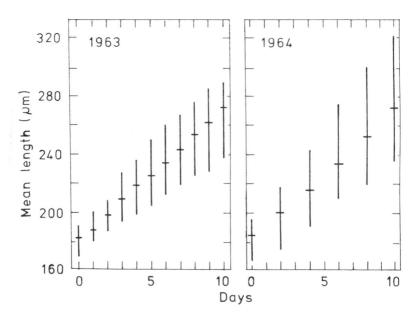

Fig. 11   The mean length of the larvae in all the large-scale experiments in 1963 and 1964. The vertical bars indicate the range of values obtained. (From Walne, 1966.)

Eight experiments were completed in 1963 and 16 in 1964 to the stage where the spat had been removed from the collectors and were large enough to be removed to the oyster ground. The average yield per experiment was 1594 and 1361 spat respectively. A number of batches had to be over-wintered in the Conwy tanks and the yield from these the following summer was 524. Part of this loss was probably because growing conditions for oysters kept in the tanks are relatively poor. The considerable deposition of silt which occurs in the winter months is also unfavourable.

Sixty experiments were completed in 1969, and these yielded on average 21 900 spat at the time of metamorphosis, but thereafter the mortality rate was very high (see Chapter 5).

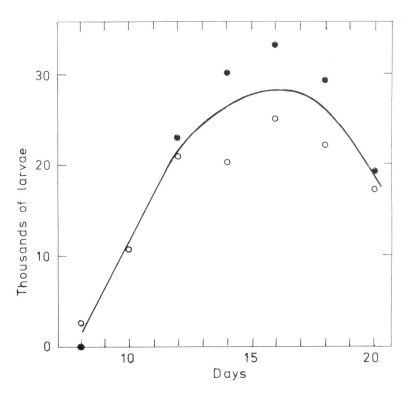

Fig. 12   The abundance of eyed larvae on successive days in the 1963 (closed circles) and the 1964 (open circles) experiments. (From Walne, 1966.)

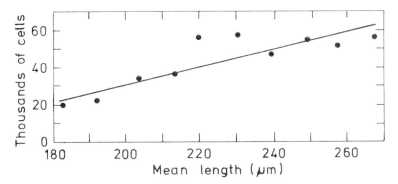

Fig. 13   The average number of *Isochrysis* cells eaten, related to the mean size of the larvae, in the 1963 experiments. (From Walne, 1966.)

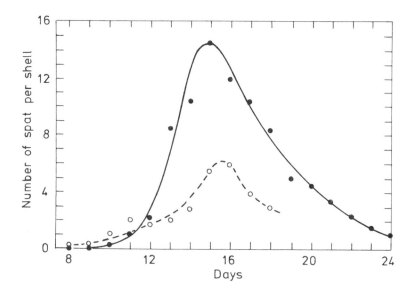

Fig. 14   The average daily spatfall on mussel shell in the 1963 (closed circles) and 1964 (open circles) experiments. (From Walne, 1966.)

Other methods for rearing large numbers of larvae

The preceding section has outlined the principles of a technique which has been found to give satisfactory results but, of course, other methods have been tried. Some of these, with their advantages and shortcomings, will now be described.

Before the war extensive use was made of cylindrical glass vessels, holding 15 to 30 litres of water, in which the contents were stirred by a glass plate slowly moving up and down through the water column. This method reached its most advanced development in the experiments at Port Erin referred to earlier, where some individual tests yielded up to 16 000 spat. The same principle has been tested at Conwy in rectangular fibreglass tanks holding 100 litres of sea water and using plunger-plates of perforated plastic, operated by eccentric cams driven by an electric motor. These were extremely efficient at keeping the larvae in suspension and some good results were obtained. For example, on one test with 750 000 larvae, a mean size of 284 microns was obtained in 8 days. However, losses of larvae were high; in the example quoted, only 44 per cent were surviving on the eighth day. It was suspected that the vigour of the stirring was cracking the shells of the larvae and this was confirmed in some tests which were made in 1 litre beakers. In two beakers gentle aeration was provided, while in two others perforated discs operated as plunger plates. Larvae of a mean length of 218μm were placed in the beakers and *Isochrysis* added as food. After 48 hours' growth the following results were obtained:

Method of stirring	Mean length (μm)	% cracked shells
Aeration	254	3.5
Plunger-plate	242	46.0

Clearly this type of agitation was too vigorous and further work on these lines was abandoned. It has, however, been our experience at Conwy that some degree of stirring gives better results. Aeration is simple to use but it is not satisfactory in rectangular tanks, where a single central air jet causes a slow overturn in the water with dead spots in the corners where large numbers of larvae accumulate. In our experience rectangular tanks are not satisfactory for larval culture. The good results obtained in laboratory experiments using 1 litre beakers suggested that vessels which are essentially cylindrical in shape, with the depth greater than the diameter, would be more suitable, and it is from this approach that the use of the polythene bins described in the previous section developed.

It can be easily seen that the swimming behaviour of healthy larvae is to alternate between periods of swimming vigorously upwards with periods of slowly sinking. This suggested that the provision of a long, vertical water column would give them a more satisfactory environment in which to develop their normal pattern of behaviour. To test this larvae have been reared in glass tubes 10cm in diameter and with three depths of water (80, 140 and 220cm) and holding 7, 14 and 21 litres respectively. A slow aeration from a jet held close to the bottom gives a very thorough but gentle turbulence. Stirring is essential, because otherwise the larvae tend to congregate close to the surface film. The average length in microns of five batches of larvae reared in these conditions was:

Days from start of experiment	Depth of water (cm)		
	80	140	220
2	204	206	205
4	223	229	229
6	237	246	249
8	253	259	264
10	263	269	274

This trend of faster growth in the deeper tubes was also shown in the percentage of eyed larvae which had developed eye-spots on certain days:

Days from start of experiment	Depth of water (cm)		
	80	140	220
8	4.8	7.2	13.5
10	17.4	26.2	31.0

Similar tubes holding 12 litres have also been used to look at the possibility of increasing the density at which larvae are cultured. The potential advantage of this is that the less water that is required for a given number of larvae, then the greater are the possibilities of using relatively expensive methods of controlling the environment. There are, however, clear disadvantages because if the density of the larvae exceeds about 1000 to 1500 per litre then their grazing activity will seriously deplete the food reserves overnight. It is therefore necessary, if rapid growth rates are required, to adopt some method of continuously adding additional food.

The development of small peristaltic pumps in recent years makes it possible to dose only a few hundred ml of culture over a 24 hour period. The procedure adopted was to estimate the number of cells which would be removed by the larvae in 24 hours. The required amount of algal culture, diluted to some extent, is then placed in a small glass reservoir and the pump speed adjusted so that the reservoir will be emptied overnight. The amount of culture required increases steadily as the larvae grow, and the high concentration involved causes a rapid accumulation of faeces and other debris. For this reason the water is changed daily. The following example, which outlines the course of one experiment, shows that the technique can be very successful.

Day	Number of larvae in 12 litres	Average shell length (μm)	Number of cells of Isochrysis added per μl in preceding 24 hours
0	680 000	192.0	—
1	529 200	193.4	0
2	—	198.2	941
3	483 000	206.0	761
4	469 000	215.6	—
5	—	222.0	1 980
6	461 000	237.0	1 153
7	531 000	248.8	1 182
8	472 000	252.1	980
9	432 000	266.3	1 200
10	480 000	269.5	1 500
11	408 000	282.1	1 379
12	464 000	280.6	1 095
13	460 000	284.7	1 422
14	428 000	285.6	1 339
15	412 000	289.0	1 428
16	416 000	291.8	—

Four per cent of these larvae were eyed on day 14, and by day 16, when they were transferred to a larger bin, 62 per cent were eyed; a heavy spatfall started on day 17. From this and similar experiments it is apparent that larvae can be reared satisfactorily at concentrations up to 40 000 per litre.

A serious technical problem with this type of culture is to accurately control the addition of food, since a slight misjudgment can lead to considerable under- or over-feeding. On occasion an excessive build-up of food has occurred when the larvae failed to start feeding for a few hours after changing the water. Further development of this technique requires equipment that would continually monitor the concentration of the food, which would be added automatically as required. The potential advantage of the method is that relatively small volumes are employed and hence very high degrees of control, and if necessary expensive measures, can be exerted.

This failure of larvae to feed after changing has been noticed on a number of occasions in a variety of rearing experiments. It does not seem to be related to the way in which the larvae have been handled, but is perhaps a measure of some quality in the water which causes the larvae to remain closed. This could include not only a toxic quality in the new sea water but also the shock of the change from the medium to which they had become accustomed. On some occasions it has been so catastrophic that formerly healthy larvae have never recovered. Such extreme examples are probably due to bacterial infection (see p. 91).

SELECTED REFERENCES

BRUCE, J R, M KNIGHT and M W PARKE, 1940. The rearing of oyster larvae on an algal diet. *J. mar. biol. Ass. U.K.*, **24**, 337-374.

LYMAN, J and R H FLEMING, 1940. Composition of sea water. *J Mar Res.*, **3**, 134-146.

WALNE, P R, 1956. Experimental rearing of the larvae of *Ostrea edulis* L. in the laboratory. *Fishery Invest., Lond.*, Ser. 2, **20 (9)**, 23 pp.

WALNE, P R, 1964. Sea-water supply system in a shellfish-culture laboratory. *Res. Rep. U.S. Fish Wildl. Serv.*, No. **63**, 155-159.

WALNE, P R, 1966. Experiments in the large-scale culture of the larvae of *Ostrea edulis* L. *Fishery Invest., Lond.*, Ser. 2, **25 (4)**, 53 pp.

4 Observations on the larvae of *Ostrea edulis*

Throughout the time that oyster larvae have been cultured at Conwy, they have been studied under a wide variety of conditions, ranging from vessels in the laboratory containing only a few millilitres of water to the largest tanks holding 400 000 litres. In this chapter a summary is presented of some of the more important facets of this work. It is necessarily incomplete and the reader will find that a number of topics which might have been expected are not discussed at all. This will be either because such lines have been tried and have so far been found to be of minor importance to larval culture in the conditions at Conwy, or because to date we have been unable to examine them in the depth that they warrant.

FOOD
When the experiments were first started at Conwy it was hardly realized that an adequate supply of the correct food was required; larvae removed from natural sea water contain many different kinds of particles in the gut. In 1923 a tank experiment, to which organic enrichment had been added, gave favourable results. In 1924 histological examination of larvae fed on starch grains showed that they had failed to ingest any measuring more than 10 microns in diameter. These two facts together initiated the long search for suitable food particles from sea water. The use of enrichment in the outdoor tanks was soon shown to be difficult to control due to the necessity of preventing the growth of fixed plants and the development of an excessive pH (see also Chapter 2, p. 33). Similarly, it was quickly realized that suitable food particles were likely to be found amongst the minute unicellular forms which either swim or float freely in sea water. Many are pigmented like plants, while others are colourless and rely on organic matter dissolved in the water for their nourishment. Sketches of some of the forms mentioned in this chapter are shown in Figure 15. For some years attention was given to the possibility of using *Fucus* antherozoids, and the first successful culture of larvae

through to metamorphosis was obtained with this food in 1930. A great variety of foods, including yeast, minced algae and *Ulva* chloroplasts, were also tried. At the same time techniques for the isolation and culture of unialgal cultures of the small single-celled forms were being developed at Conwy and elsewhere. At first these produced cultures of the more robust and non-motile forms, such as *Chlorella* and *Coccomyxa*. Pure cultures of suitable marine flagellates were first available in 1936 and eventually examples from many algal classes were obtained and tested.

In 1934 a start was made on amplifying the work at Conwy by detailed studies at Port Erin on the food requirements of larvae. At both stations the use of plunger jars holding 15 to 30 litres became established as the preferred method for laboratory rearing. The disadvantage of the method was the labour required to look after such large containers and the provision of sufficient algal culture. For this reason the post-war development of a technique whereby larvae could be reared, in good numbers, to metamorphosis in 1 litre glass beakers greatly helped the laboratory investigations.

Initially in most rearing experiments the larvae were reared through to metamorphosis, and, as rearing was restricted to the summer months, only a few experiments could be made each year. Subsequently experiments lasting for all periods from 24 hours onwards have been used; however, I am now of the opinion that a day is too short, since some of the growth will be a reflection of the previous history of the larvae. Moreover, there is a statistical error of ± 3 to 5 microns in measuring the mean length of 100 larvae and because this is a significant proportion of an average growth increment of 10 microns in 24 hours, it is difficult to obtain statistically significant results in this time. At the time of writing we usually run the experiments either for 4 days or until metamorphosis. Large-scale experiments, such as those described in the previous chapter, provide supplies of larvae of various sizes. In general the food preferences are similar throughout the larval phase, although some current work suggests that some algae, or combinations of algae, can affect metamorphosis.

In much of the earlier work the lack of food species known to be good food for use as a control was a severe handicap. The only way to proceed was to test a food

Plate 9   The polythene bins used for the large-scale culture of bivalve larvae.

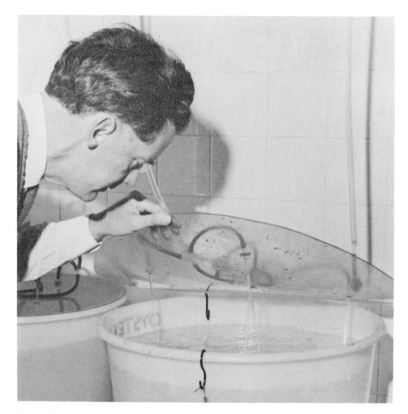

Plate 10   A culture of oyster larvae. Note the five aerators which reach to the bottom of the vessel. The black hook holds a line attached to the plastic sheet laid on the bottom of the bin for the spat.

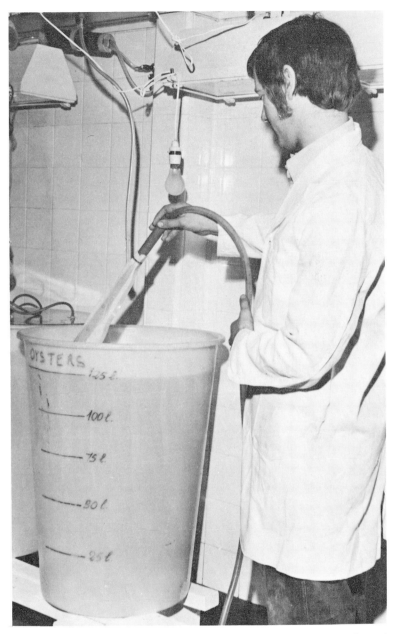

Plate 11   Changing the water. The contents of the bin will be siphoned
out through the hose with the larvae being retained by the mesh sleeve
over the inlet.

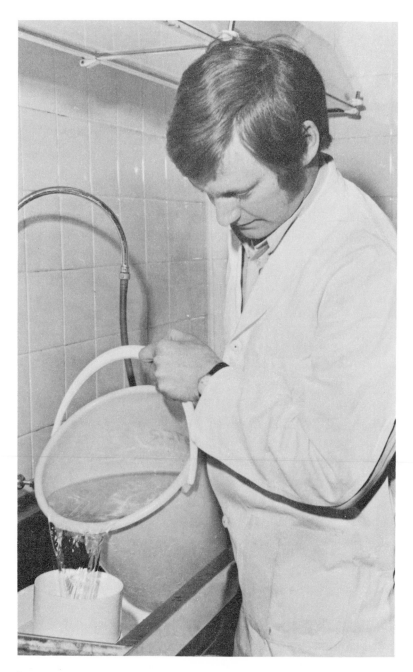

Plate 12   Larvae being transferred to a small sieve before washing and replacing in a clean bin.

organism repeatedly, and if it never gave good results then it could be assumed to be unsatisfactory. Other species gave promising results, with occasional lapses, and these became the subject of more comprehensive investigations. The use of the radioactive isotope ^{32}P proved useful at this stage. If algae are grown in a phosphate-deficient medium to which some ^{32}P has been added, most of the radioactivity is quickly incorporated into the cells. If these are then fed to the larvae, a proportion of the ^{32}P is assimilated and built up into larval tissues; by measuring the amount taken up under various conditions, an estimate is quickly obtained of the rate of assimilation. This was found to be closely related to the growth of the larvae.

No food other than unicellular algae has been found satisfactory, and within this group of plants different species vary from a complete food to ones which support no growth at all. There is a tendency for a particular systematic group, comprising related species and genera, to be either poor or good foods, and the following paragraphs summarize the results obtained from tests with representatives of various orders.

The distribution between the orders is somewhat unequal because of the rather stringent requirements which have to be satisfied before tests can be carried out. It may sound like stating the obvious, but first the plant has to be available in uni-algal culture with a knowledge of a medium and cultural condition in which it will form dense cultures. It is the last 20 years that have seen a very considerable expansion in the isolation and subsequent taxonomic description of these minute forms. A few species have been isolated at Conwy but generally we have relied on algologists working in other laboratories. Many of the species which have been isolated do not form thick cultures; it is best to be able to use as a test species one for which the addition of 1 per cent of culture to sea water is equivalent to adding 1 per cent of an *Isochrysis* culture containing 10 000 cells per μl. Other species do not remain suspended and are not then available for the larvae. Some species, which are typically found only in the cooler months of the year or in the open ocean, do not grow at the temperature used for culture of the larvae. A further problem which has still to be tackled is the variation in the chemical composition of the algal cells which it is known can occur when the culture conditions are varied. Are new vigorously growing cells better or worse food than cells which are older, usually larger, and packed with food reserves? Do variations in the medium, often based on sea water, affect the composition of the food? Do other sources of variability, such as light or temperature, also have their effect?

A further point which requires emphasis is that we have little knowledge of the food of oyster larvae in nature. Although clearly there is successful development of larvae in the semi-natural Conwy tanks or in the Essex rivers, we have little information on the species composition of the natural flora on which the larvae feed in those conditions. Those species which have been tested in the laboratory may well be quite rare in the sea, but are able to withstand and grow in the artificial environment of the laboratory.

So far most of the tests of food have taken either growth over fairly short periods or the production of spat as the criterion of success. It would be more informative if we had a knowledge of the biochemical changes occurring within the larvae, but their very small size makes quantitative measurements difficult. The meat content of a mature larva is about one microgramme and large numbers are required for chemical analysis. Of course poor foods are obvious, but we are now reaching the stage where more subtle effects are appearing. It is very probable that the food value of some algae varies according to the culture conditions, and we have some indications that the vigour of the spat may depend to some extent on the food received by the larvae. Do some feeding regimes result in larvae which are unhealthily fat, while others promote a good growth of protein but are too lean for the larvae to lay down adequate food reserves to tide them over metamorphosis? Whether the larvae require vitamins and other trace substances in solution in the water is at present unknown.

Chlorophyceae

Unicellular green forms are among the commonest plants which grow when samples of sea water are enriched with salts and allowed to stand. These species, which include motile and non-motile genera, are very hardy in culture and readily produce dense growth. For these reasons considerable attention has been paid to this group, but unhappily with indifferent results. Early experiments with species of *Coccomyxa* and *Chlorella* gave little or no larval growth and it was suggested that their rather thick cell wall could not be attacked by the digestive system of the larvae. In a series of nine experiments with *Chlorella stigmatophora* some growth occurred in only two, and similar results were obtained with *C. marina*. The related minute form *Nannochloris atomus*, which is only 2-3μm in diameter and therefore perhaps more readily digested, did produce spat in two experiments out of nine; in some of the others it is probable that the cell density was too low.

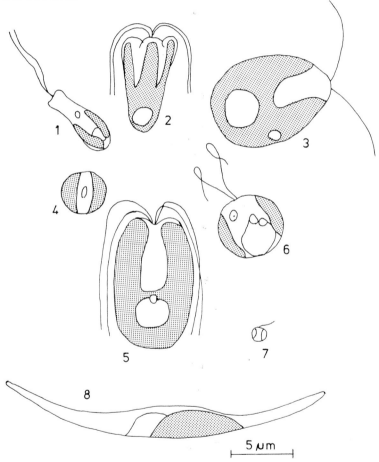

Fig. 15 Some algae which are commonly used in oyster culture (all on the same scale). The chromatophores are shown shaded. 1, *Isochrysis galbana*; 2, *Pyramimonas grossii*; 3, *Dunaliella tertiolecta*; 4, *Chlorella stigmatophora*; 5, *Tetraselmis suecica*; 6, *Dicrateria inornata*; 7, *Micromonas pusilla*; 8, *Phaeodactylum tricornutum*.

The motile species *Dunaliella tertiolecta* (often known in the earlier literature as *Chlamydomonas* III, Figure 15) has given some promising results when bacteria-free cultures were tested, but the results were more erratic than in the controls. These were experiments which only lasted a day, and from subsequent experiments of longer duration on spat it is most likely that this species would not support growth for a longer period.

Prasinophyceae
The recent founding of this group by systematic algologists is of great interest to oyster biologists, since it contains two genera *Pyramimonas* and *Tetraselmis* (Figure 15) which were formerly members of the Chlorophyceae, a position which appeared to be anomalous because their food value was high in comparison with most members of that group. *Pyramimonas grossi* was used as long ago as 1936 (as flagellate H), when it showed some promise; later it gave much better results when a series of four experiments all produced spat. Unfortunately it is not very easy to grow and it will not be a useful food until satisfactory culture conditions have been worked out. In recent years a related species, *Pyramimonas obovata* has been found to grow well in 20 litre flasks and it is a good food for larvae. Several species of *Tetraselmis* (also known as *Platymonas*) have given excellent results. A long series of comparative experiments has shown that the growth of larvae fed at 10 cells per μl is as good with *T. suecica* as with *Isochrysis* at 50 cells per μl — approximately equal volumes of food. This species grows very well in our standard culture conditions (see p. 55), and because of this has become established as one of our standard larval foods.

A third species in this group is the minute form *Micromonas pusilla*. Cultures are yellow-green in colour, although the individual cells, which are only about 1.5μm in size, appear nearly colourless under the microscope. Plants of this or closely related species are very common in samples of inshore and offshore waters. In the Conwy breeding tanks, where the water is free from silt, and hence observation is easier, flagellates less than 2μm in size are usually the most abundant plants. Some good results have been obtained with this species but because of its small size high cell densities need to be maintained.

Haptophyceae

This group contains several genera of flagellates, yellow-brown in colour, which were formerly classified with the *Chrysophyceae* and are well-established as good food organisms. The most notable is *Isochrysis galbana*, which the reader of Chapter 3 will have seen has been our standard larval food for many years. *Dicrateria inornata* (as flagellate B) and *D. gilva* (as flagellate C) are similar in value, although more difficult to culture.

Several species of *Chrysochromulina* have been tested and all have shown some promise, but they are not easy to grow in thick cultures. Mention should be made here of the remarkable complexity of structure of this group which has been demonstrated in recent studies with the electron microscope (eg. see Parke, Manton and Clarke, 1955).

The remaining species of interest is *Prymnesium parvum;* this is a widespread form and blooms of it are known to have caused mortalities of fish. In an experiment with oyster larvae fed with 20-56 cells per μl, all the larvae were dead within 7 days and no growth occurred; the control fed with *Isochrysis* went on to produce spat. On one occasion a bloom of *P. parvum* developed in one of the outdoor rearing tanks and here again all swimming larvae soon disappeared.

Chrysophyceae

This group includes *Chromulina pleiades* (as flagellate C) which has given very good results, up to half the larvae metamorphosing, but unfortunately cultures of this species no longer exist. The closely related *Monochrysis lutheri* has been used extensively in larval culture work, particularly at Millport (Scotland) and Milford (USA). We find that growth is often not as good as that obtained with *Isochrysis*, although some experiments with bacteria-free cultures of *M. lutheri* gave similar results to the *Isochrysis* controls.

Cryptophyceae

This is a colourful group of flagellates including green, brown and red forms. Many do not grow very well in culture but some success has been achieved with *Cryptochrysis rubens* and *Cryptomonas acuta*. *Hemiselmis rufescens* was used with limited success at Port Erin, as was the very similar *H. virescens* in bacteria-free culture at Conwy.

Cyanophyceae

The minute (1μm) species *Synechococcus elongatus* has been isolated from the Conwy breeding tanks but in a series of three experiments no larval growth was obtained.

Bacillariophyceae

This group comprises the diatoms. They are an abundant group in the sea and are characterized by the presence of a siliceous skeleton. Most of them are too large to be eaten by oyster larvae. Some tests with *Phaeodactylum tricornutum* gave irregular and unsatisfactory results, but from the results of experiments in which this plant was fed to spat it is unlikely that the growth rate obtained would have continued for long. Trials with two relatively small diatoms, *Chaetoceros calcitrans* from Japan and *Cyclotella nana* from North America, have given some excellent results, although they are difficult to maintain in large-scale dense culture at all times.

Mixed algal foods

It is remarkable that a single algal species can provide an adequate diet and it is not surprising that recent work has demonstrated that better results can be obtained with mixed diets. A mixture of any two of the following three species — *Isochrysis galbana*, *Tetraselmis suecica* and the diatom *Chaetoceros calcitrans* — will give larvae which are significantly larger and contain a greater proportion eyed, after 8 days, than any of the foods on their own. In addition more spat are obtained from larvae which have been fed a mixed diet.

The experiments have also revealed another important aspect of using mixed foods: those spat which were derived from larvae fed on mixed diets grew faster and had a lower mortality rate than those from larvae fed on single foods, although all had received the same diet after metamorphosis. It appears that the food eaten by the larvae can have a marked influence on the vitality of the spat, and clearly this has important implications for commercial practice. In some places it may be practicable to obtain a nutritionally satisfactory mixture by only coarsely filtering the sea water so that the natural food supplements that provided by culture. This is, however, an uncertain procedure and it is preferable to have a more strictly controlled diet.

Other foods

So far we have discussed the use of photosynthetic algae as food, but other nutritious particles have also been considered. Colourless flagellates are often abundant in the sea, and some have been shown to be useful in Japan, from where a strain was obtained in 1953. This grew vigorously in sea water enriched with starch, which encouraged the development of a dense population of bacteria on which the flagellate fed. The population rose to about a thousand per μl within 4 days, but it was found in practice to be difficult to avoid adding some of the media to the larval culture. This, with the bacteria, caused a growth of bacteria in the larval culture which was inimical to their development. Yeasts have also been examined. Baker's yeast was tried very early on and marine species isolated from sea water have also been tested; neither was successful. Dried liver, which had been powdered sufficiently fine for the larvae to swallow, yielded promising results, but this and other non-living foods all suffer from the difficulty that uneaten food contaminates the water. In some cases they may also be difficult to keep in suspension.

Another source of food, which is not strictly artificial, is the normal algal food which has been preserved in some way. It would be very convenient to produce algal cultures throughout the year and to 'bank' supplies against peak demand and as an insurance against the failure of cultures in critical periods. Some success has been reported from America, in feeding dried and freeze-dried unicellular algae to clam larvae. We have been able to test a spray-dried *Chlorella* from Japan, a culture of *Monochrysis* vacuum-dried in mannitol from Millport, and cultures of our own *Isochrysis* which were freeze-dried for us at the British Museum. Although these all gave reasonable suspensions, with a high proportion of single cells, none of them gave significant growth of oyster larvae. The discovery of a good preserved food, even if only to use in emergencies or to 'top-up' at times of peak demand, would be a most valuable addition to hatchery procedure.

Concentration of food

Not only must the food presented to the larvae possess an adequate food value, but it must be present in suitable concentrations in the water. This was examined in some detail using algae 'marked' with the radioactive isotope ^{32}P.

The combined results of seven series of experiments are shown in Figure 16, where the relative radioactivity of the larvae after about 22 hours is plotted against the density of

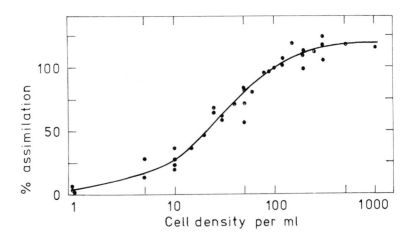

Fig. 16   The combined results of a number of experiments measuring the uptake of ^{32}P by larvae fed at various densities of *Isochrysis galbana*. (From Walne, 1965.)

cells of *Isochrysis* at which they have been feeding. The value obtained when feeding in water containing 100 cells per μl was taken as 100 and the values obtained at other cell densities adjusted accordingly. It is clear from this figure that variations in the cell density between 10 and 100 cells per μl have a considerable influence on the amount assimilated. At a density of 50 cells per μl, about 70 per cent of the maximum possible assimilation is obtained. This was achieved at about 300 cells per μl but it is not practicable to rear larvae at these densities because there is a considerable production of pseudo-faeces, and rapid fouling of the culture takes place. Similar experiments made with six other species of algae with different sizes of cells gave similar curves, but the position of the inflexion varied a good deal for the different species. A convenient way of comparing species was to calculate the food level at which an increase in the cell density by 5 cells per μl increased the ^{32}P assimilated by 2 per cent. For *Isochrysis* this corresponds to 58 cells per μl, a density at which larval assimilation is about 75 per cent of the maximum. With a small species such as *Micromonas minutus* the equivalent level is at 132 cells per μl, while at the other extreme, for *Dunaliella tertiolecta*, it is 25 cells per μl.

The thick-walled, non-motile *Chlorella stigmatophora* was also tested and this gave a quite different result. Larvae were fed at various cell densities between 25 and 150 cells per μl, but no systematic difference in assimilation was obtained throughout the series. This species is not a good food (see p. 82), and apparently the larvae are unable to assimilate ^{32}P from this alga in the normal manner. This confirms the suggestion that the digestive enzymes cannot penetrate the cell wall and the animal is unable to mechanically rupture the cell.

The previous paragraphs have referred to assimilation but a number of tests have shown how closely this is related to growth. To obtain clearly defined curves the larvae need to be grown for several days in the density under test. An illustration of this was shown in one series of experiments where larvae were taken directly from the large-scale rearing bins and little difference in growth rate was obtained over the subsequent 48 hours between those fed at 30, 60, 90 and 120 cells per μl. They had all been well fed while in the bins and their reserves were probably sufficient to maintain a high growth rate (in terms of increase of shell length) when food was relatively scarce.

Quantity of food eaten

A number of estimates have been made of the number of cells of *Isochrysis* eaten per day. The longest series of results was obtained from the large-scale rearing experiments in 1963. In these standard experiments in 75 litre bins, the cell density was estimated daily and the size and number of larvae every 2 days. From these data the average number of cells eaten per day has been calculated and is shown graphically in Figure 13; it increases from about 20 000 cells per day when the larvae are first liberated to 60 000 as they approach metamorphosis.

A similar estimate was made in 32 experiments using *Isochrysis* marked with [32]P. In this case the minimum number of cells which the larvae must have eaten was calculated by dividing the radioactivity in the larvae by the known activity in the cells (Figure 17). Since the digestive efficiency of the larvae is not known, this calculation is a minimum figure, but it is notable that the slope of this and

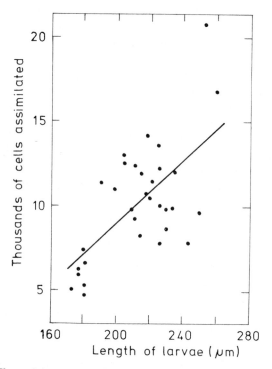

Fig. 17   The minimum number of cells assimilated by larvae in 24 hours when feeding on *Isochrysis* at a concentration of 100 cells per μl. (From Walne, 1965.)

the previous curve are very similar; in both the food consumption increases by a factor of about three during the larval stage. This technique estimated that small larvae will assimilate 5000-10 000 cells of *Isochrysis* per day. In comparison, they will assimilate 16 000-23 000 of a small species such as *Micromonas*, but only 400-1200 of a large species like *Dunaliella tertiolecta*.

Chemical composition of the food

Miss D M Collyer in a study at Burnham-on-Crouch of the chemical composition of eight species of unicellular algae, made some analyses of the large-scale cultures at Conwy. As we are interested in using these plants as food, it is most useful to consider their composition in terms of weight of compound in a cell, and in Table 4 the results of some of the analyses are summarized in this manner. Five samples were examined from the large-scale, semi-continuous *Isochrysis* cultures which varied from those only 5 days old, which had not been harvested before, to those of 56 days, which had been harvested 30 times.

A further series of analyses made in 1969 of *Isochrysis* confirmed these results. These two sets of data show how even a standard culture method produces algae with considerable differences in composition with the result that our standard practice of feeding 50 cells per μl of *Isochrysis* to the larval cultures can lead to considerable differences in food concentration. On the basis of the analytical data 'starch as glucose' can vary between 25 and 595μg per litre, protein between 50 and 610μg per litre, and the algal volume between 2900 and 5500μ^3 per μl. The effect of these variations on the growth of larvae and spat is worthy of further exploration.

The main result of these analyses of *Isochrysis* and other species, some of which are also shown in Table 4, was that the differences in composition between the species were no greater than the differences caused within a species by changes in the medium or in the time since subculturing. The reason why one plant is a good food for oyster larvae and another is not has not yet been found out with certainty for any food. Is it the shortage of some trace constituent in one food, resistance to digestive enzymes, inability of the larvae to rupture the cells, or some other unsuspected cause?

One interesting index that was calculated from these data was the 'cell density index', obtained by dividing the dry weight by the volume of packed cells. Cell density did not always parallel cell volume, and apparently some species, such as *Dunaliella*, often produced a very heavy cell whereas *Isochrysis* tended to produce a rather light watery cell.

LIGHT
At the present time there is no clear evidence that growing larvae are influenced by light. Cole and Knight-Jones (1949) reported a rather random variation in vertical distribution in the Conwy tanks, although tow-net samples in the Helford river suggested that perhaps in bright sunshine the larvae congregated in the lower levels. A number of laboratory experiments at Conwy comparing growth in the dark with normal laboratory lighting showed no advantage with either condition. It is, however, suspected that in certain conditions

light can be of importance. For example, on a number of occasions larvae have been water-changed in the normal way and placed in 2-litre graduated cylinders. It is then not uncommon for the larvae to remain for some time very close to the bottom. It might be suspected that the new water was distasteful, but some tests showed that if a bright light is then shone on to the larvae they will, on occasions, quickly rise from the bottom and start to swim vigorously within a few minutes of the stimulus starting. This phenomenon cannot always be demonstrated, and its application to the large-scale experiments in the glass tubes referred to on p. 72 did not lead to enhanced growth.

As metamorphosis approaches, however, the situation is different; here light becomes an important stimulus, affecting both the development of the larvae and their behaviour in seeking for suitable settlement surfaces. In an early experiment frosted glass plates, ground side down, were suspended in the large outdoor tanks during settlement; two plates were clear glass and two had their backs coated with black paint. They were hung at mid-depth and each day the spat were counted and removed; over a period of 18 days 238 settled on the clear plates, compared with 735 on the dark ones. In a parallel experiment in 1947 similar plates, which were clear, white and black, were tested for 12 days hung at the surface and the bottom. The numbers of spat recorded were as follows:

Background colour of plates

	Clear	*White*	*Black*
Surface	20	50	75
Bottom	132	159	457

An interesting variation of this experiment was made using glass funnels instead of glass plates. Two large funnels with the neck uppermost were suspended in a tank with a clean oyster shell, concave side down, held over the neck. It was shown in 1938 that this arrangement concentrated the vertically swimming larvae since a shell in this position caught 97 spat in 5 days, compared with none on a control alongside. In 1947 one of the funnels was painted with black paint, and the other was left clear; over a period of 6 days the shell over the dark funnel caught 23 spat, compared with 14 on the shell over the clear funnel. This difference is small, and the reason became clear when the funnels were examined. Twelve spat were found to have settled on the glass in the clear funnel but in contrast the dark one carried 1611 spat. The experiment was repeated in 1948, but the long necks of the funnels were removed. In 6 days the shell over the clear funnel received 365 spat against 581 on the shell over the dark funnel.

From these experiments we may conclude that in the clear still waters of the Conwy tanks, mature larvae seek dark areas in which to settle. Further confirmation of this was obtained in 1939 when settlement was compared on clear and dark plates during daylight and at night. Over a period of three nights 21 spat settled both on the clear and the black plates, but during the daylight periods 27 and 131 spat settled respectively. As expected, the spat did not discriminate between the light and dark surfaces at night.

Some evidence has also been obtained that the rate of settlement is reduced at night. In 1939 groups of oyster shells were hung 9 inches, 3 feet, and $5\frac{1}{4}$ feet below the surface, and the settlement was recorded during 7-hour periods for six consecutive days. The average spatfall per shell was as follows:

	Daylight		*Darkness*
	0630-1330 hours	*1430-2130 hours*	*2230-0530 hours*
Surface	88.1	107.0	30.6
Mid-depth	9.1	8.7	3.5
Bottom	2.4	0.6	3.5

At first, the standard technique for rearing larvae in the hatchery did not call for light and satisfactory spatfalls were obtained. However, some experiments showed that in the presence of continuous light spatfall was both earlier and greater in quantity. For example, a batch of larvae was reared in a 75-litre bin in darkness for 9 days and then split into two; at this time their mean size was 268μm and none had developed eye spots. One half of the batch were continued in the dark while the other half received continuous light from a 60W bulb. The spatfall was recorded daily on two strings of mussel shells hung near the surface, and on frosted glass plates laid on the bottom; these were examined at irregular intervals. Spatfall began on the fourteenth day in the lit bin, but not until the nineteenth day in darkness (see Figure 18). The total spatfall recorded was:

	Light	Dark
10 shells at surface	4661	989
10 plates on bottom	9598	5698

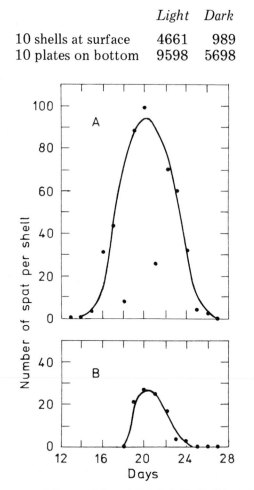

Fig. 18   The mean daily spatfall on mussel shells hanging at the surface in experiments (A) cultured in the light, and (B) cultured in the dark. (From Walne, 1966.)

A more detailed series of factorial experiments showed that some light in the 24 hours before settlement had a stimulating effect, but its intensity should not exceed that used during the settlement period. During the latter, light intensities at the water surface greater than 600 lux were favourable for settlement. This is illustrated in the following table which shows the number of spat which settled on 20 mussel shells in 24 hours in 4-litre vessels exposed to various light intensities:

Light intensity in the settlement period	Light intensity in the preparatory period		
	100 lux	500 lux	1000 lux
100 lux	42	109	20
500 lux	297	441	97
1000 lux	432	419	439

All these experiments indicate that light has an important effect on larvae at or near the time of settlement, but much remains to be done. Nothing is known about the effect of light of different wavelengths, or of diurnal fluctuations compared with continuous light, nor should it be forgotten that the light is operating in a system containing photo-synthetic algae — the food of the larvae. It could be that the light is operating through the algae, but until we have a non-living food this point will be difficult to resolve.

TEMPERATURE

Breeding populations of flat oysters are found where the temperature of the water in summer is usually in the range 16-26°C, although there is evidence that some populations will breed in temperatures as low as 13°C. It is difficult to observe the effect of temperature on larval growth in the outdoor tanks and in the field, since only occasionally is there a sufficiently sharp peak in the population for the growth rate to be accurately estimated. Some detailed laboratory studies have been made and the following paragraphs review these results.

The effect of temperature on larvae both before and after liberation is summarized in Figure 19. The period before liberation was determined by opening a number of oysters until a gravid female whose eggs were in the early cleavage stage was found. These eggs were then cultured in glass vessels kept at different temperatures. The developing embryos were examined periodically and the time taken to reach a shell length of 160μm was determined; this point was used because it was known that, given the opportunity, larvae will start to feed at that stage. As 170-180μm is the usual size at liberation a further one to two days should be added to

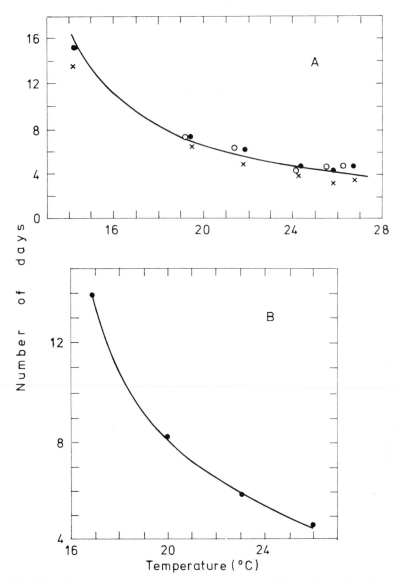

Fig. 19   The relation between temperature and (A) the time taken for three broods of oyster larvae to grow from early cleavage to a mean shell length of 160μm, and (B) the time taken for larvae to grow from 175μm to 250μm. (From Walne, 1965.)

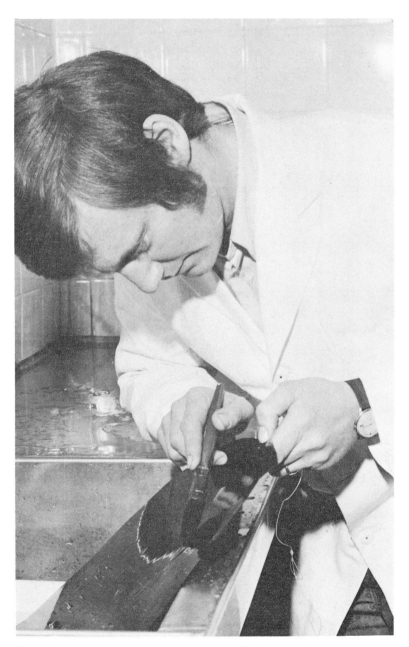

Plate 13   Stripping oyster spat within 24 hours of metamorphosis from a PVC sheet.

Plate 14   Oyster spat in a re-circulation system in the laboratory.

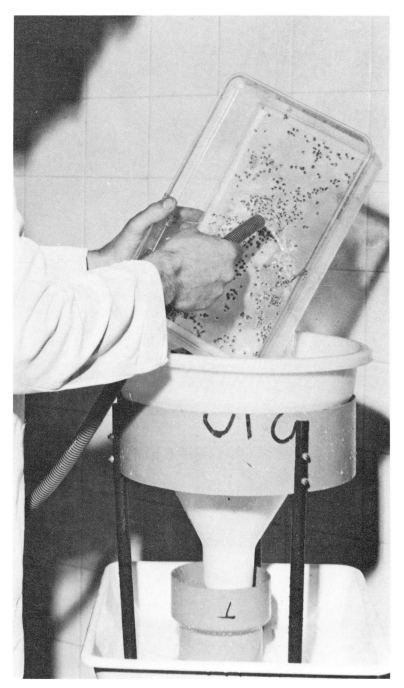

Plate 15   Washing oyster spat before transferring to clean water.

Plate 16   Oyster and clam spat in flowing sea water.

give the full period of development within the mantle cavity.

The growth of the free-swimming larvae at various temperatures when fed with 100 cells μl of *Isochrysis* is summarized in Figure 19b. In this case the time taken is that to grow from 175μm (the size at liberation) to 250μm; when the population reaches this mean size some of the faster-growing animals are starting to become eyed. The curves drawn in Figure 19 correspond to the formula $x.y = K$, where x is the time in days to reach 160 or 250μm, y is the temperature in °C above biological zero (10°C for the early embryos and 13°C for the larvae), and K is respectively 66.6 and 58.1. Biological zero is the point at which growth no longer occurs. Since there is a physiological difference between the two stages (the embryos are not feeding and the larvae are), it may well be that there are real differences in the temperature of no growth in the two stages.

The effect of temperature on feeding activity has been studied mainly by allowing larvae to graze on *Isochrysis* containing the isotope [32] P. It was found that if two different cell densities were fed at a variety of temperatures the ratio of assimilation at any temperature was constant. For example, in one experiment larvae were fed at cell densities of 10 and 100 cells per μl at temperatures in the range 16.4-27.9°C. Throughout this range the uptake at 10 cells per μl was constantly about 25 per cent of that at 100 cells per μl.

From a number of similar experiments it has been possible to build up a table of indices of assimilation (Table 5). Here the assimilation of food has been taken to be 100 at 20°C and a food density of 100 cells of *Isochrysis* per μl. The other figures show the change in food uptake which will occur if the density of food or the temperature are varied either together or separately. Since growth and assimilation are closely linked, this table provides an indication of the growth that will be obtained.

In summary, oyster larvae can be reared at a wide variety of temperatures. Good results have been obtained at Conwy in the range 14-25°C; usually we maintain our experiments at about 22°C. Higher temperatures, which will mean faster growth, are convenient if small vessels standing in a water bath are used, but it should be remembered that larger aerated vessels standing in a controlled temperature room are usually about 2°C cooler than the air temperature.

BACTERIA

It was observed in some experiments in 1954 that bacteria became sufficiently numerous for fluffy white colonies to be seen on the sides of the beaker. This stimulated enquiry into the density of the bacterial flora which develops when oyster larvae are reared in small glass vessels. It was soon found that apparently the greater the bacterial population, the fewer the number of spat obtained. This result could have arisen if the bacteria were affecting the growth of the larvae or, alternatively, if the water quality was such that when it stimulated the growth of bacteria, then it also had a depressing effect on the development of oyster larvae.

The next step was to rear the larvae in an environment where the bacterial density was low. It is fairly easy to free sea water of its bacterial population, but less easy to produce both the food and the larvae in a bacteria-free condition although both have now been done by workers at Millport. At the temperature of 20-22°C used for larval rearing, marine bacteria divide frequently, and if only a few gain access to the experiment a dense population soon builds up. A further complication is that if most, but not all, of the bacteria in a culture vessel are removed, then the numbers which are present after 24 to 48 hours are usually greater than those in a vessel in which the number at the beginning had been less drastically reduced. Presumably the balance of species has been so disturbed that the remaining few are able to flourish unchecked. For these reasons the use of antibiotics was introduced to control the bacteria without otherwise affecting the larval cultures.

The first experiments were made with the sodium salt of penicillin G. This antibiotic was comparatively cheap and had been shown by other workers that at suitable concentrations it did not harm the development of fish embryos and marine flagellates. Samples from 15 broods of larvae were cultured through to spatfall in 1-litre beakers, to which sufficient penicillin was added to give a concentration of 50 international units per ml at the beginning of the experiment. Spat were collected on mussel shells and it was found that substantially more (average 306.8 per experiment) were obtained from those larvae which had penicillin than those which did not (average 103.3 per experiment). At the rearing temperature and at the pH of sea water, penicillin will only remain active for a few days. The water was not changed in these experiments and it was found that the bacterial population was only controlled at a low level by the penicillin for the first few days. It then rose, often to a higher level than in the controls, and here additional penicillin had no further controlling effect.

In further experiments penicillin and streptomycin were used together in order to obtain control of a wider spectrum of bacteria. This was achieved and in some, but not all, series of tests better spatfall was obtained in the presence of mixed antibiotics than with either on their own. Based on these data it has become the standard practice to add 1ml of a stock solution of antibiotics to each litre of sea water used for larval cultures. The stock solution is prepared by dissolving in distilled water 15g of the sodium salt of penicillin G with a potency of 1670 u/mg and 33.5g of streptomycin sulphate with a potency of 745 u/mg, and making up to 500ml.

Subsequent experiments have confirmed these results and have shown that not only is the spatfall increased but also that the rate of growth is enhanced in the presence of antibiotics. A number of other antibiotics at various concentrations have also been tried and three, Ilotycin, Colomycin and Ceporin, give results which are as good or better than the standard mixture. As they are at present more expensive it is unlikely that they will supplant the present mixture, although it is useful to have alternatives to combat those bacteria which are not affected by it.

Unfortunately the use of antibiotics does not confer immunity from bacterial disease. In July 1969 we had a period when it was impossible to rear oyster larvae. Freshly established cultures would collapse within one or two days, and older cultures, where the larvae were growing vigorously, could collapse within a few hours. Typically, nearly all the larvae would stop swimming and collect together on the bottom; microscopic examination showed that decomposition of the larval tissues was well advanced.

We were able to isolate strains of bacteria from such cultures and tests with oyster larvae showed that some had toxic properties. The effect during a test could be dramatic: a small loopful of a pathogenic culture placed in a small tubeful of swimming larvae would result in a cessation of swimming within a few hours, and after 24 hours the larval tissues were almost completely decomposed. Unfortunately it has not been possible to distinguish the strains from those which are commonly found in the laboratory's sea water supply and therefore we have been unable to trace their origins. This knowledge is clearly important for hatchery management and also leads to speculation on the importance of such outbreaks in the sea. In our cultures bacteria can come with the sea water, with the algal food, with the larvae themselves, or from the air. Usually the sea water has several hundred bacteria per ml and the addition of 10ml per litre of *Isochrysis* increases the population of bacteria by several thousand per ml. The larvae will also bring some bacteria, although this is reduced by rinsing with chlorine solution at the water change.

FOULED SURFACES AND THE GREGARIOUS FACTOR

An accidental observation in 1938 started a series of investigations into the influence of other organisms already present on the surface on the settlement of oyster larvae. Some shells had been left undisturbed in one of the breeding tanks for a few days and when taken out they were found to bear a great many recently settled spat and a few older ones; similar shells which had been cleaned daily had caught only a few spat. This suggested that settling larvae were attracted either by the fouling of the undisturbed surface or by the presence of some older spat; subsequent inquiry showed that both these factors are important.

As soon as a surface is immersed in the sea, organisms become fixed to it. Bacteria and diatoms are early colonizers and they, with other varieties of algae, will form a slimy film over the surface within a day or so in the summer months. This has been shown to be important in encouraging the settlement of a number of species, for example serpulid worms, bryozoa and some barnacles.

Its influence on the settlement of oysters was examined in the breeding tanks by offering shells which had been fouled for various periods of time. In the first experiment in 1939 ten pairs of matched shells were placed in a breeding tank. The shells were examined daily and one set of ten was cleansed by thorough brushing; the other set had all spat removed with a needle but the fouling film was left undisturbed. No significant settlement took place for the first 18 days that the shells were exposed, but in the next 6 days 1264 spat settled on the fouled shells compared with 831 on the controls. The experiment was repeated in 1940, when the two batches at first caught about the same number of spat, but in the second fortnight 4340 settled on the fouled shells compared with 3731 on the controls. This suggests either that 2 to 3 weeks' fouling was required to be effective, or, alternatively, that the gradual increase in the number of scars left where spat have been removed was the deciding factor.

These alternatives were explored in 1940-41, when matched sets of 10 shells were first exposed to fouling before being offered for oyster settlement. In both years batches of shells were hung in a net bag for about 3 weeks in the Menai Straits. When they were brought back to Conwy they were well covered with a bacterial and algal film, and ascidians, filamentous algae, patches of polyzoans, and *Pomatoceros* were very common. In both years the shells were placed in the breeding tank when spatfall was in progress, and the numbers of spat which set in the first 24 hours were as follows:

	1940	1941
Fouled shells	495	2144
Clean control shells	113	614

These differences were kept up for many days but the results quoted here, which are for the first day of the test, showed clearly that it was the period of soaking in sea water and the accumulation of fouling that was important in the previous experiments and not the accumulation of scars of young oyster spat.

There still remains the possibility that it is not the presence of organisms that is important but some more subtle, perhaps chemical, effect which results from the prolonged soaking. This possibility is not yet fully eliminated, although an experiment in 1941 in which the settlement was observed on shells from which the encrusting organisms had been removed on one half, suggests that it is the fouling that is important. These shells caught 1766 spat on the fouled half compared with 871 on the cleaned half in 14 days. Control shells caught less spat (231), but it was difficult to clean a fouled shell to the same standard and this may have affected the result. Alternatively the proximity of the heavily fouled surface may have been important.

In an experiment in 1947 6 heavily fouled shells were divided into two; one set of three was placed on a tray in the breeding tank, but the other set was not placed in the tank until the macroscopic fouling organisms had been removed. Six clean control shells were also added. The shells were examined daily and all spat removed; the total recorded, per shell over 11 days, was as follows:

Heavily fouled (three shells)	518 per shell
Partly cleaned (three shells)	441 per shell
Clean (six shells)	22 per shell

This suggests that the main effect is due to the fouling film rather than to the larger organisms.

The second aspect of fouling investigated was whether the presence of the same species, oyster spat, had an attractive influence on settling larvae. Ten pairs of matched shells were placed, concave side down, on a wooden slatted tray hung in the breeding tank. One shell of each pair was left undisturbed during the experiment, whereas the spat on the control was removed daily with the point of a needle. In this way a similar degree of fouling was built up on all the shells with the exception of the experimental set, which acquired a gradually increasing population of oyster spat. In 1939 during exposure to settlement for 13 days the 10 experimental shells caught 2670 spat compared with 1265 on the controls; in 1940 the respective figures were 2388 and 950, during 7 days' settlement. The records show that each day the experimental shells caught between two and three times as many spat as the controls. Apparently a few spat on the shells were as effective in attracting others as many spat, although this was to some extent confused in both years by the fact that at the time when the shells carried a large number of spat (200-400), the intensity of setting was light. As the control shells were receiving 20-200 spat each per day, the general level of discrimination may have been reduced; it is also possible that there is some exclusion effect if the shell is already heavily covered with spat.

Although the ratio between the experimental and control shells was similar during a range of setting intensities, it was noticeable that on any particular day the spatfall on a shell tended to be proportional to the number of spat already on it. This could have been due to the increased attraction of the larger number of spat, but it must be remembered that the shell with more spat must have been more attractive than its fellows on the preceding days, and what we are seeing here is that some shells may themselves be more or less attractive than their fellows.

These experiments strongly suggest that the effect is obtained by the larvae encountering the spat during their crawling phase, rather than by smelling them from afar, since the 'smell' — if any — of oysters must pervade the whole tank when it is remembered that there were about 600 stock oysters present. In an experiment in which the spatfall on shells grouped round two adult oysters was compared with controls, less spat settled on those nearest the oysters, presumably due to interference by the feeding current, but smaller oysters can make a slate surface attractive. Two slates were hung for 7 days in the breeding tanks. To one slate six one-year-old spat were attached with marine glue; the control only had the marine glue. The newly settled spat were counted and removed daily. The total recorded was 982 on the experimental slate, of which 427 settled on the older spat, while only 100 settled on the control slate. Similarly the presence of previously settled spat confers an attractive quality to the settlement surface in laboratory experiments. Small slate panels that carried spat were kept for 2 to 4 days until the young oysters had grown to a size of $400\text{-}600\mu$m; the controls were similar panels without spat which were soaked in sea water for the same period. Sets of both panels were placed in glass vessels holding 4 litres of sea water and a number of mature larvae. The following table shows the results from three experiments:

	Mean number of spat on plates	*Average number of larvae settling in 24 hours*	
		With previous spat	*Control*
1	7.5	14.8	1.7
2	19.0	21.0	5.5
3	25.0	54.5	13.0

In each case four or more spat settled on those panels carrying growing oysters for each one which settled on the control. The degree of fouling from micro-organisms must have been similar on both sets of panels, but presumably the crawling larvae encountered the growing spat and were encouraged to settle in their vicinity. But was it the smell which was attractive, or was it simply the presence of an irregularity on an otherwise smooth, featureless surface that had a stimulatory effect?

Fouling of spat collectors, whether by an algal and bacterial film, or by oyster spat, is an important factor in the operation of the large breeding tanks. Let us suppose that the tanks have been filled in the usual way and a number of spat collectors added. After a month or six weeks an intense spatfall begins, and it is judged necessary to spread this over further collectors which are then added. However, the collectors which have been in the tank for some time have acquired a film of fouling and perhaps a few spat as well; these will continue to be very attractive and spat will settle on them two or three times as abundantly as on the new collectors. The remedy is not clear. If all the old collectors were removed when the new ones were added, then the disturbance to the layer of detritus and algae on the tank floor would be so great that the spatfall would come to a premature end. Lighter collectors, suspended in the tank, would allow those with a full complement of spat to be removed without causing a general disturbance.

Following the early work discussed in the preceding paragraphs, it has been demonstrated that the gregarious phenomenon is shown by a number of other sedentary marine animals, notably some fixed species of marine worms, and by barnacles. Barnacle larvae are not only attracted by already settled barnacles, but are also attracted to surfaces treated with an aqueous extract of barnacle flesh (Crisp and Meadows, 1962). Settling oyster larvae are also attracted by similar extracts painted on to collectors. A suitable extract can be prepared by homogenizing the meat from an oyster in about 100ml of sea water and then clarifying by first coarsely filtering and then centrifuging. To test the efficacy of the extract it was painted on to frosted glass plates which were then dried either in the laboratory or in a cool oven. When settlement was about to begin, 10 to 12 plates, some treated and some controls, were distributed on the bottom of the standard 75 litre rearing bins. The plates were removed after 1, 2 or 3 days and the number of spat counted. The average ratio of 17 sets of observations showed that the spatfall was 3.58 times heavier on the treated plates. It is tempting to assume that this is the same phenomenon as that found when shells bearing spat a few days old were tested, but there are important differences. On the one hand, the crawling larva comes into contact with a surface filmed with some components of an oyster's flesh and, on the other hand, it encounters the shell of an oyster which may be days, months, or years old. Whatever the cause, the technique is sufficiently useful for it to be used as a standard procedure in our hatchery technique.

The activity of plates coated with extract has been shown to be reduced by a wide variety of treatments: a temperature of 300°C, sodium hypochlorite and caustic soda are particularly effective. Plates can, however, be stored, after drying, for several weeks and there is considerable latitude in the strength of the prepared extracts. Extracts prepared by homogenizing 1g of oyster flesh in 10ml of sea water contain, on average, 1.45mg of protein per ml, and this can be diluted at least tenfold before the attractiveness of the treated plates declines. The stock extract should be stored in a refrigerator, since some loss of activity has been found if it stands at temperatures over 20°C for as short a period as 16 hours.

A comparison of extracts made from various tissues and adjusted to contain the same amount of protein showed that the adductor muscle and gills yielded an extract as favourable as one prepared from the whole animal, but that from the mantle was below average. An interesting point is that the extra-pallial fluid is nearly as active as the standard extract. Since this fluid will have a simpler chemical composition than the extract of the body tissues, this could be a point of attack for further investigation into the chemical nature of the extract.

The extract is apparently fairly specific. Tests on the settling larvae of *Ostrea edulis* have been made with extracts prepared in the usual way from *Ostrea lutaria*, *Crassostrea gigas*, *C. angulata* and *Mytilus edulis*. None of these induced a greater settlement than that obtained on the untreated control plates, with the exception of *Ostrea lutaria* which was only a little less effective than *O. edulis*. This suggests that members of the same genus may stimulate each other's larvae to settle.

What is the chemical nature of the components of the extract which are attractive to larvae at the time of metamorphosis? Ultra-filtration and dialysis in Visking tubing (24A mean pore size) have shown that the active components of the extract could not pass through, and were therefore of fairly large molecular size. A precipitate was obtained if 2.4 molar ammonium sulphate was added to the standard extract. This precipitate, which contained about 75 per cent of the protein in the extract, could be dissolved in a standard buffer solution. After dialysis to remove the ammonium sulphate, it was found that this purified extract contained about 70 per cent of the activity of the original extract. It is less stable to room temperature than the original extract, but treated plates can be dried without loss of activity.

The reader will have seen that we have three phenomena involving living organisms which apparently influence larvae in their search for a suitable place for metamorphosis: fouling organisms, other oyster spat, and an extract of oyster meat. The question now arises, is the latter a laboratory phenomenon or does it bear some relation to what happens in the sea? Is the 'smell' which we apply to our test panels similar to that found by the crawling larva when it encounters a shell of the same species, or is it carried by the exhalent current? What are the physical characteristics of surfaces which are favoured by larvae? Our ability to rear large numbers of larvae in the laboratory is an important step forward towards detailed studies on larval behaviour. Similarly, our ability to increase the attractive nature of settling surfaces is an important tool in the large-scale production of oyster spat.

SELECTED REFERENCES

BAYNE, B L, 1969. The gregarious behaviour of the larvae of *Ostrea edulis* L. at settlement. *J. mar. biol. Ass. U.K.*, **49**, 327-356.

COLE, H A and E W KNIGHT JONES, 1939. Some observations and experiments on the setting behaviour of larvae of *Ostrea edulis*. *J. Cons. perm. int. Explor. Mer*, **14**, 86-105.

COLE, H A, and E W KNIGHT JONES, 1949. The setting behaviour of the larvae of the European flat oyster *Ostrea edulis* L., and its influence on methods of cultivation and spat collection. *Fishery Invest., Lond.*, Ser. 2, 17, ,No. 3, 39 pp.

KNIGHT JONES, E W and P R WALNE, 1951. *Chromulina pusilla* Butcher, a dominant member of the ultraplankton. *Nature, Lond.*, **167**, 445-446.

MILLAR, R H and J M SCOTT, 1967. Bacteria-free culture of oyster larvae. *Nature, Lond.*, **216**, 1139-1140.

WALNE, P R, 1958. The importance of bacteria in laboratory experiments on rearing the larvae of *Ostrea edulis* (L.). *J. mar. biol. Ass. U.K.*, **37**, 415-425.

WALNE, P R, 1963. Observations on the food value of seven species of algae to the larvae of *Ostrea edulis*. 1. Feeding experiments. *J. mar. biol. Ass. U.K.*, **43**, 767-784.

WALNE, P R, 1965. Observations on the influence of food supply and temperature on the feeding and growth of the larvae of *Ostrea edulis* L. *Fishery Invest., Lond.*, Ser. 2, **24 (1)**, 45 pp.

5 Oyster spat

The stage in the life history of marine bivalves which is colloquially called 'spat' cannot be precisely defined, but it is a convenient term for the period from metamorphosis until they reach a few millimetres in size. Among fishermen it covers the period until the beginning of growth in the second summer. The use of the term leads to the word 'spatfall' which refers to the period of metamorphosis.

Although spat are better at withstanding unfavourable conditions than are the larvae, they are more sensitive than the adult stages, and their successful culture requires care. They are an attractive stage in the development of the oyster since they are transparent, and the functioning of the gills and heart can be readily observed under the microscope in an intact animal which has metamorphosed on to a glass plate.

Although strictly we are concerned with very small animals, it is convenient to discuss in this chapter some relevant aspects of bivalve biology which are more readily observable in larger animals but which are, nevertheless, relevant to spat culture. The first part of this chapter discusses the effect of various factors on the growth and survival of oyster spat, and finally the methods which have been found satisfactory for spat culture at Conwy are described.

DENSITY ON SURFACE

A strong batch of larvae can give an intensive spatfall over a period of a few days, and unless the larvae are kept at a very low concentration, dense settlements can occur on the collectors. It appears that there is little tendency for the larvae to space themselves out. From this the question arises, do the spat interfere with each other before they reach the size at which they are touching and overgrowing each other? This was examined by obtaining a dense settlement of spat on glass plates 9cm square, and then obtaining even distributions at various densities by removing the surplus.

The average growth and survival obtained was as follows:

Spat per cm^2	After 14 days	
	Mean length (mm)	% survival
4.0	3.3	100
2.4	2.8	77
1.0	2.9	90
0.44	3.3	96
0.11	3.2	80

From this it appears that very substantial densities of small spat can be grown for a short period on relatively restricted surfaces if there is an adequate water supply; if a million spat could be evenly distributed on both sides they could be grown on a piece of sheet 2.3m square (about 16 square feet) for 14 days after metamorphosis.

Even at this small size, oyster spat filter appreciable quantities of water and the effect of crowding too many into a tank soon becomes noticeable. For example, in one test different numbers of newly-metamorphosed spat were cultured in 1 litre beakers for 3 weeks; the water was changed three times a week. At the end the mean size of the spat was:

Initial number of spat per litre	Mean size after 21 days (mm)
10	5.3
50	3.2
100	2.3
300	2.4
500	1.5

The decrease in growth rate with increasing numbers of spat is closely related to the rate at which the spat were feeding, and hence exhausting the food supply in the water (see also Figure 21). The result is very probably due to starvation rather than other undesirable effects like the accumulation of waste products.

FEEDING RATE

Within a few hours of metamorphosis the gill bars can be seen to be growing out into the mantle cavity, and after 48 hours the gills are well developed and functional. The rate at which spat remove their food from the water has been studied by keeping small glass plates with spat attached in jars of filtered water, to which cultures of the flagellate *Isochrysis* have been added. As the animals feed, the clarity of the water gradually increases and this can be measured in such instruments as 'Harvey' photometer (Harvey, 1948). This instrument, using a 15cm tube, will measure changes of 2 or 3 cells per μl and from these data a theoretical concept, 'the volume swept clear', can be calculated. This volume is the minimum amount of water that the oyster must have filtered free of all particles to achieve the observed degree of clarity. Of course, the animal might have swept twice the volume with only half the efficiency. This type of measurement does not tell us, but since we are interested in feeding, the point is unimportant; we are concerned with how much has actually been eaten. Two points have merged in these studies: the rapid increase in the filtering rate within the first few weeks of life, and the way in which the filtering rate increases as the food becomes less plentiful in the water. The first point is illustrated by the graph in Figure 20, which relates the filtering rate of young oysters to their diameter. These data were obtained by observing the rate of removal of *Isochrysis*, and from them it can be calculated that 1000 spat about 4 weeks old (5mm in size) will filter 7 litres (1.5 gallons) per hour at 21°C, and within another 4 weeks (15mm in size) this will have risen to 50 litres (11 gallons) per hour. Similar examples are given for quahogs in Chapter 6. Another way of using these data is to calculate the volume of water in which oyster spat will have to be kept in order to obtain a certain reduction in the food concentration overnight. For example, reference to Figure 21 shows that 50 per cent of the food will be eaten in 24 hours if 100 spat measuring 2.8mm and filtering at the rate of 2ml per hour are kept in 50 litres of water. To avoid excessive reduction in the food supply one of two alternative courses has to be adopted: a sufficient volume of water, or feeding more often than once every 24 hours.

The second point is that a reduction in the concentration of food is compensated to some extent by spat increasing the

114

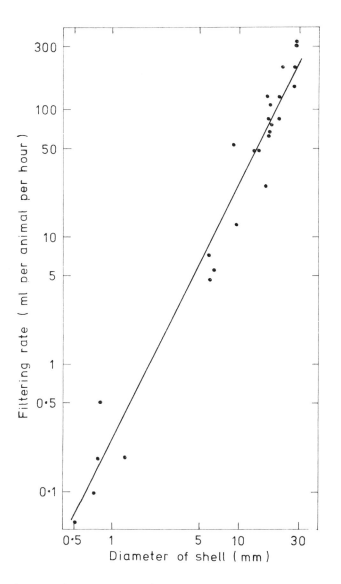

Fig. 20   The filtering rate of *Ostrea edulis* spat related to shell diameter when feeding on *Isochrysis* at a temperature of 21°C.

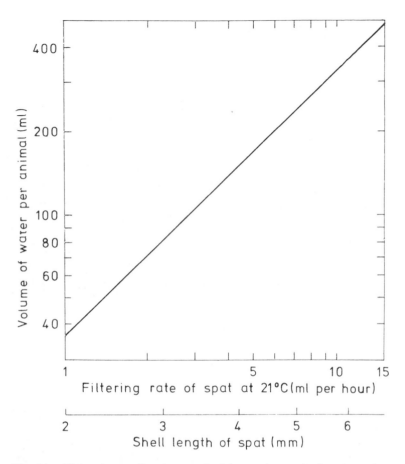

Fig. 21 The volume of water required by oyster spat of various sizes to give a 50% reduction in food concentration in 24 hours.

rate at which they filter. This is a common phenomenon with filter-feeding animals and has the result that an insufficient food supply is exhausted as rapidly as an adequate supply, but as an increased amount of energy has to be used to obtain the food, the growth rate is reduced.

It will be shown later that oyster spat grow well when kept in filtered sea water enriched with 50 cells of *Isochrysis* per μl. Since a 5mm spat will remove all the cells from 7ml per hour, it consumes 350 000 cells in this time. Assuming that a culture is produced with a cell concentration of 10 000 per μl (see p. 57), the daily ration of a 5mm spat is

Plate 17   Frames used to protect quahog and palourde spat when planted on the beach.

Plate 18   Carrying tiles bearing oyster spat down the beach at Tal-y-foel.

Plate 19   Tiles just after placing on racks at low water mark.

Plate 20   Tiles in August after 12 months in the sea.

118

Plate 21   Experimental tray with batches of young oyster spat.

Plate 22   Tray with 18-month-old oyster spat.

Plate 23   Tray with experimental oysters.

Plate 24   Plot of Pacific oysters at low water mark.

contained in 0.84ml of algal culture. By the time it has reached 15mm in size, the daily ration is about seven times as large. In practice we find that excellent and more economical results can be obtained by growing spat in raw sea water so that they can utilize the natural food present, which is supplemented with smaller amounts of algal culture.

EFFICIENCY OF FILTRATION

The mechanism by which bivalves remove their food is complex and contains various methods by which the animal can vary its filtration efficiency between pumping considerable volumes of water, and yet removing very few particles, through an increasing degree of retention until many particles of the size of bacteria are retained. A full description of these methods, which involve changes in the size of the opening of the pores in the gills, alterations to the ciliary beat, and variations in the flow of mucus over the gill surface, will be found in Gareth Owen (1966).

Our interest in the efficiency of retention has arisen from our studies on food requirements; how much of the apparent variation in the food value of different species of algae is due to the inability of oysters to catch the food? The problem has been approached in two ways: by observing the feeding rate on algae of different sizes and, secondly, by comparing the particulate content of water before and after it has passed through the oyster.

In the former method, which was used on a number of 18-month-old oysters, the animals were fed successively on four or five different species of algae. Oysters were placed individually in 1 litre beakers containing filtered water enriched with one species of alga. The theoretical filtration rate could then be estimated from observations on the rate of removal of the algal cells. At the end of the test the oysters were transferred to beakers containing another species of alga. In this way the filtering rate of the same animal on foods of different sizes could be compared. Any one animal was used in only four or five trials, so care was taken to use always one species of alga (*Phaeodactylum*) as a reference. This was fed to every oyster and provides a standard by which all the foods can be compared. The following table shows the average filtration rate on various foods (at least 10 oysters were used in each trial), expressed as a percentage of that given by the same oysters when fed on *Phaeodactylum*.

	Per cent
Neutral red	5.9
Aquadag	18.4
Dag 312	12.0
Nannochloris atomus	41.1
Isochrysis galbana	13.8
Chlorella stigmatophora	37.0
Dunaliella tertiolecta	35.1
Tetraselmis suecica	148.0
Prascinocladus sp.	114.0

This list is in order of increasing particle size and demonstrates how filtration apparently increased as the test materials used increased from a solution of a vital stain (neutral red) through small graphite particles and the smaller algae, and finally reaching cells similar in size to those of *Phaeodactylum*. This method of experimentation has certain disadvantages; one cannot be certain that the animals are going to behave in a similar manner when placed in the different suspensions. The small species may have also been less palatable ones. A further disadvantage is that fairly high concentrations had to be used in order to get a response on the photometer.

It is possible to collect the water from the exhalent current by fitting a rubber sleeve over this part of the animal and connecting it directly to an overflow. This method has been widely used in studies on the rate of pumping by oysters (see Galtsoff, 1964). It can give excellent results, but there is some uncertainty about their validity, because the animal is somewhat confined and very precise levelling is required to ensure that water is not being forced through the mantle cavity. Tests with apparatus of this type at Conwy showed that small oysters would frequently remove 90-100 per cent of *Phaeodactylum* during its passage through the mantle cavity when pumping 2 or 3 litres per hour, but the same oysters would generally remove only 60-80 per cent of *Isochrysis*.

FOOD

The number and size of particles which are eaten by oyster spat has been discussed and now we should consider their nature. In the same way as for larvae, the only satisfactory food is unicellular algae and also, again as with larvae, the species vary considerably in their food value.

A considerable number of algal species have been tested both on their own and in combination. A suitable method of handling the oyster spat in a trial was to obtain a spatfall on small (9cm × 9cm) frosted glass plates by placing these on the bottom of the standard larval rearing bins. The plates were usually painted with an aqueous extract of oyster meat which was allowed to dry before being offered to the mature larvae. The test was started with 2 or 3 days of metamorphosis, and the number of spat was then reduced to 20-25 per plate. The spat could be measured while attached to the plate and the increase in mean size of the group over a 3-week period was a good indication of the value of the food offered. For example, Figure 22 shows how the growth rate soon changes if the spat are transferred from a good to a poor food and *vice versa*. As shown earlier in this chapter small oysters filter

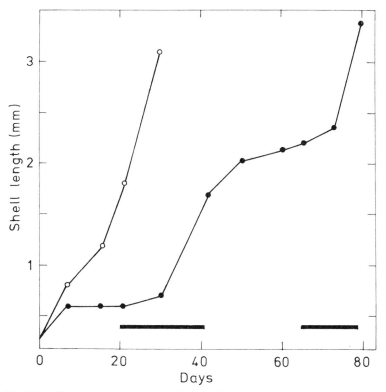

Fig. 22   The growth of oyster (*O. edulis*) spat when fed on *Isochrysis* (open circles) or *Dunaliella* (closed circles). The *Isochrysis* was fed to the latter for two periods, indicated by the horizontal bars.

an appreciable volume of water each day, and the volume of water in the experiments had to be so arranged that the spat did not eat more than about 10 per cent of the available food in a day.

It soon became clear that not only was the nature of the food important but so was its abundance in the water. An example is provided in Figure 23 which shows the average growth made by oyster spat kept in various densities of the flagellate *Isochrysis galbana* for three weeks after metamorphosis. From this result it is apparent that an adequate supply of food has to be present in the water, but too much is as harmful as too little. It is noticeable that spat are more

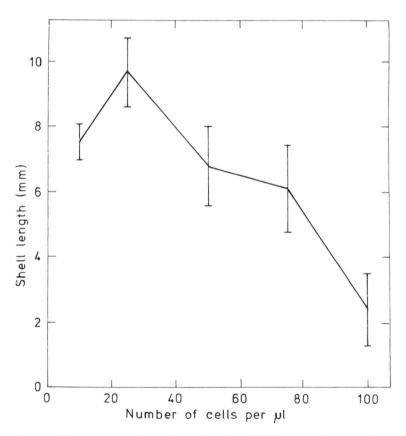

Fig. 23   The average size of oyster (*O. edulis*) spat 47 days after settlement when fed on various densities of *Isochrysis*. The vertical bars indicate the 95% confidence limits.

sensitive to an abundance of food than are the larvae (compare Figure 16); to attempt to provide a good rich feed can be deleterious.

From this it follows that it is necessary to test each new algal food at several concentrations to ensure the use of the density which gives the maximum possible growth rate. In each series of tests with a batch of spat, a control, usually *Isochrysis*, was also used. The various algal species can be ranked in order of food value by comparing the mean size achieved by the spat in the most favourable concentration of the species under test with the size achieved in the corresponding control. The reliability of this method was shown when indices calculated in this manner for a given

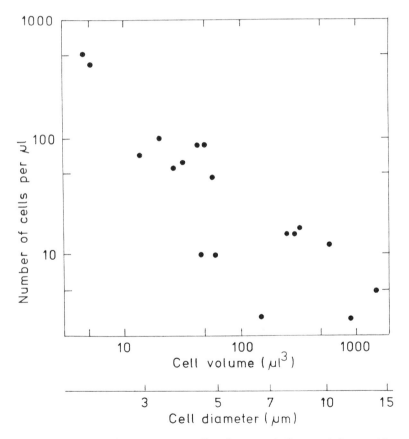

Fig. 24   The relation between cell volume and the most favourable density of algal food for oyster spat. (From Walne, 1970.)

food were compared with trials with different batches of spat made at widely separated times. The indices for 18 species of algae are listed in Table 6. Five species of algae supported growth rates as good as or better than *Isochrysis*. Three are flagellates: *Dicrateria*, *Tetraselmis* and *Monochrysis*, and two are diatoms: *Skeletonema* and *Chaetoceros*. It is very noticeable that all of these are known to be good food for larvae, with the exception of *Skeletonema* which has not been tested. Those species which are poor food for larvae, *Dunaliella*, *Chlamydomonas* and *Chlorella*, are also of little value for juveniles, and it appears that there is little change at metamorphosis in the ability of oysters to thrive on a given species.

The optimum density of the food in the water varies according to the size of the species concerned. This is illustrated in Figure 24 where the average volume of the cells of each of the species listed in Table 6 is plotted against the abundance of cells which gave the best growth. This increased from about 4 cells per μl with the largest cells tested to about 500 per μl for the smallest. As might be expected it is the overall volume of material obtained in the feeding current which is important, and presumably larger amounts lead to the oyster's filtering system becoming clogged.

MIXED FOODS
Although excellent growth has been obtained with some foods over relatively short periods it is very likely that trials of longer duration will show up serious shortcomings in some of them. There is no reason to suppose that any one algal species will contain the complete nutritional requirements in the correct balance, and this view is confirmed in experiments which have been made on mixed diets. No doubt some foods are bad because they cannot be digested and therefore cannot be improved by being used in a mixture, while others may have some very minor component missing which could be supplied by small amounts of another food. At the time of writing it is doubtful if this latter case has been found. It is much more usual for two reasonably good foods to be improved by being fed together.

One way of tackling the problem has been to use unfiltered sea water. This usually contains a considerable assortment of species which might be expected to supply all the necessary micro-nutrients, although at most times the total volume of algal material will be lower than that found to be optimal in the spat feeding experiments. Considerably enhanced growth has been obtained when either *Isochrysis* or *Tetraselmis* was added as an enrichment to sea water which had only been strained through a fine mesh (about 70μm) to remove the larger organisms, compared with the same food used in fully filtered water. The amount of growth obtained varied from time to time due to the vigour of different batches of spat and to changes in the food content of the water. In a series of experiments spat were grown for a period of 21 days after metamorphosis in filtered and unfiltered sea water enriched with either *Tetraselmis* or *Isochrysis*. The average length (mm) of shell of the spat after this time were

Food	*Water*	
	Fully filtered	*Strained*
Isochrysis	4.48	8.17
Tetraselmis	5.50	9.82

Clearly strained water leads to much better growth, and it is desirable to be able to reproduce these conditions at will rather than having to rely on the vagaries of the natural flora.

The problem of examining various food mixtures is very considerable, since there is no underlying theory which enables us to say that if species A is used then some of species B should also be used to remedy the deficiencies of A. It is not too laborious to test a mixture of two foods, but care has to be taken that the total volume of the algal food present is the same in each mixture.

A satisfactory experimental design is given by growing six samples of the same batch of spat on a diet ranging 100 per cent of species A through gradually increasing proportions of B until 100 per cent is reached. This 'block' of six trials can then be repeated at another level of total food expressed as μm^3 of algae per μl of water. When a blend of three species of food is tested a quite simple experimental design will require 18 tests and yet this mixture will still be a very simple one compared with the natural flora of the sea.

A considerable number of mixtures have been tested. These have included blends of two good foods, good and poor foods, and two poor foods. None of them have improved a poor food. So far as we know at present a poor food for spat cannot be improved by the addition of a small amount of good food. An illustration of this is provided by the following table showing the result of feeding spat for 2 weeks on a *Tetraselmis/Monochrysis* mixture.

Percentage (by volume) of Monochrysis	*Shell length after 21 days (mm)*
100	1.1
80	1.4
60	1.6
40	1.8
20	2.0
0	2.2

The results obtained when a mixture of two good foods are used have been variable. On occasions clear improvement has been obtained as is shown by the following result obtained by feeding *Isochrysis* and *Tetraselmis*:

Percentage (by volume) of Isochrysis	*Shell length after 21 days (mm)*
100	5.0
80	7.3
60	7.3
40	6.7
20	4.2
0	4.6

However, on other occasions the same mixture has shown that one or other of the species is better than the other. Whether this is due to variation in the chemical composition of the algae (see p. 67), in extra-cellular products present in the media, in the bacterial flora or to quite unsuspected causes is unknown. It is not impossible that different batches of spat, originating perhaps from different strains of oysters, have diverse requirements.

LIGHT

Direct sunlight is harmful to oyster spat. In one experiment sea water flowed through a shallow trough which was divided into three compartments. The first and the last were open to the sun, while the centre compartment was so arranged that the contents were in the dark. Glass plates bearing oyster spat of average size 2.8mm were placed in each division in the middle of July. Four weeks later the average size of the spat was 13.2mm in the dark compartment but only 8.8mm in each of those exposed to the sun; there was little difference in mortality between the three compartments. The period of the experiment was not particularly sunny; the daily average bright sun in the 4 weeks was 3.8, 2.8, 5.8 and 4.2 hours respectively.

In another test in the first 3 weeks of April two troughs, one light and one dark, were provided with running water warmed to about 15°C. Three different batches of oyster spat were used, with initial sizes of 0.6, 2.6 and 3.3mm. After 3 weeks those in the dark trough measured 2.1, 5.7 and 7.1mm respectively; those in the light trough were all dead.

Dim light probably does little harm but we now make it a standard practice to cover all the outdoor tanks containing young spat. This has the additional advantage that the growth of attached filamentous seaweeds is stopped. These can be very vigorous in the summer months when they will quickly obstruct the flow of water. They are difficult to remove and the trouble caused by covers is well outweighed by their advantages.

A further aspect of the influence of light which needs more study is the fact that oysters of all ages are very sensitive to small changes in light intensity. Open, feeding, animals will quickly close at the passage of a shadow and if this happens frequently it is possible that it would eventually affect the growth rate.

TEMPERATURE

As the temperature of the water is raised so the growth rate of the spat will increase, provided that sufficient food is supplied. The relation between temperature and growth is shown in Figure 25 for three species of bivalve spat reared at Conwy. From these results, it appears that good growth can be obtained at 24-25°C for *Ostrea edulis* and since this is well above the water temperature in the British Isles the choice of temperature used for growing large numbers of spat will depend on a consideration of the cost of heating large volumes of water. From one point of view it is desirable to

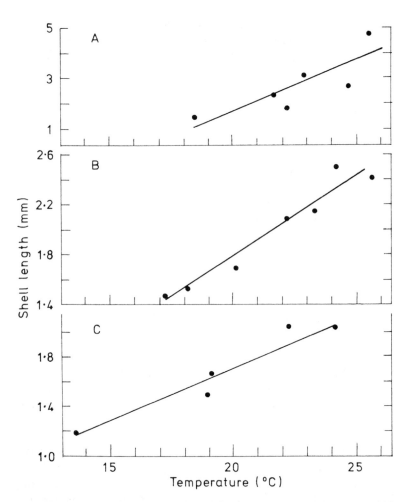

Fig. 25   The shell length of spat grown at various temperatures: (A) *Ostrea edulis* for 17 days after metamorphosis, (B) *Mercenaria mercenaria* for 28 days — initial size 1.05mm and (C) *Venerupis decussata* for 21 days — initial size 1.00mm.

reduce the time that the spat remain within the hatchery. This period depends on the time taken to reach 3-5mm in size and also on the season of the year. The following table shows some examples of the growth and survival of representative batches of spat attached to collectors which were moved from the warm, sheltered conditions inside to running sea water outside.

Month	Size (mm) when moved outside and size after about 4 weeks	Percentage mortality in about first 4 weeks outside
February	1.8	100
	3.2	100
March	2.7	100
	6.9	94
	9.7	100
April	5.4-5.6	0
	7.2-7.3	0
May	0.5	59
	3.2-8.1	3
	5.2-8.6	9
June	0.6-2.0	78
	0.6-3.1	43
July	0.7-2.7	68
August	7.7-9.8	45
	8.2-12.2	0
	10.8-12.4	19

Two important points emerge from this: mortality is usually high when spat only a few days old (less than 1mm) are moved outside, whatever the season, and mortality is always high when spat are moved outside in the winter and early spring, whatever their size. This second point presents a serious problem, since hatchery rearing can well begin in January so that the early spat can reap the advantage of a full growing season, but larvae obtained at that period will give spat 5-10mm in size by late February or March. By then they will be requiring considerable volumes of food and water, but if they are transferred to unheated water considerable mortalities will occur. It may be possible to reduce these by gradually reducing the temperature in the laboratory, so as to imitate the changes which occur in the autumn when growth is gradually reduced. Although spat grew more rapidly at high temperatures they differ physiologically from those grown in cooler water. In one experiment samples of the same batch of spat were grown from an average size of 4mm to 10-12mm in water of various temperatures and of course those in the cooler water took longer to achieve the required size. At the end of the period the meats and shells were weighed and it was found that the ratio of the two components of the body differed appreciably (Figure 26); apparently the reaction of shell and meat growth to temperature is different, and it may be that spat grown in cooler water, which have a relatively higher meat content, would be better able to withstand cooler conditions.

The spring is a time when a small degree of water warming can be valuable. The phytoplankton is often abundant in

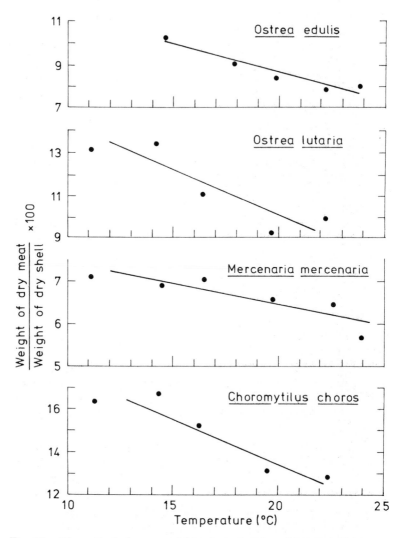

Fig. 26   The ratio between weight of meat and weight of shell in spat grown at various temperatures.

March and April, but the temperature is not quite high enough to permit growth of spat to start. Some trials in which oyster spat 2.7mm in size were transferred for 3 weeks in early April either to flowing water at ambient temperature, which ranged from 4 to 8°C, or to a warmed supply, which ranged between 11.4 and 13.0°C, showed that those in the cold water all died, but those in the warmed water survived and grew to 3.3mm. Those in the warm water were gradually raised in temperature to 15°C over the next month, during which they grew to 10mm.

Although the heat rise need not be very great, 5° perhaps, a considerable amount of heat is required to warm running sea water if substantial volumes are required. The cooling water of electricity generating stations is a potential source of a low grade heat. The temperature of the discharged water is 5-10°C above that of the sea and this could provide a considerable stimulation to growth if there is sufficient suitable food in the water. In all stations the water is chlorinated, either continuously or at intervals, to prevent the growth of sedentary organisms in culverts and pipes, and it is possible, although this point still requires testing, that this treatment has an undesirable effect on the small food organisms on which the oyster depends. This warmed water may be better used for crustacea and fish, since in this case the food is added to the tanks and the animals are not dependent on that brought in by the water.

HANDLING HATCHERY-REARED SPAT

The final paragraphs of this chapter outline the procedures which have been used at Conwy for handling fairly large numbers of oyster spat. It falls into two parts: in the first we are concerned with the method of handling spat attached to collectors, and in the second with the much newer technique of stripping the spat from the collectors when they are only a few hours old.

When the spat collectors were removed from the culture vessels they were placed in 400 litre fibreglass tanks filled with warmed sea water which has been strained through a 68 micron plastic mesh. Sufficient algal culture was added to obtain an enrichment of 2.5 cells per μl of *Tetraselmis* or 25 cells per μl of *Isochrysis*. Normally the cell density was restored to these values daily, but when large numbers of spat were present a dosing pump was used to add culture continuously. It is most important that the number of spat in a tank should be kept at a reasonable level, since overstocking will soon cause a reduced growth rate and a high mortality.

It is vital to keep the water moving in the vicinity of the spat, because local depletion of food can soon occur; the importance of water movement as a stimulant to feeding has been shown earlier (p. 22). In one arrangement that we used water ran continuously from a higher tank to a lower one and periodically a pump, controlled by a float switch, returned the water to the higher level. Where large numbers of collectors are needed a better result could be obtained by using a pump to circulate the water continuously round individual tanks. To reduce the risk of disease spreading throughout the tanks, too many tanks should not be coupled together.

The water was changed and the tanks cleaned twice a week. The use of strained rather than filtered sea water brings into the tanks many small organisms which proliferate rapidly in the warmth and good feeding conditions. To reduce a build-up of these, the collectors should be washed. This was done conveniently by standing them for a few minutes in a tank of fresh water, while theirs was being changed. A little chlorine (3 ppm) can also be used but it is doubtful if it gives a better result. Tests have shown that its use at times when growth is poor, and the spat are perhaps less resistant, can lead to a greatly increased mortality rate.

After a period of 3-6 weeks under these conditions the spat grew to several millimetres in size and required considerable volumes of food. At this stage they will do much better if they are transferred to outdoor tanks where they can receive a substantial supply of running water. The dangers of moving spat out when they are either too small or the water is too cold has been emphasized earlier in this chapter (p. 115). To avoid the risk of stagnation we have found that the best arrangement is to hang the collectors in single file in a long narrow tank into which the water enters at the bottom at one end and overflows from the top at the other.

The spat are ready for removal when they are 1-2cm in size; since they are not strongly attached to the flexible plastic collectors, they are readily removed by a combination of flexing and brushing the sheets. The shells of some of the spat are damaged during the removal and while this is being made good the spat are kept in trays with a mesh base. These trays are arranged in a larger tank and water flows continuously into the top of them, past the animals and out through the base and so to waste. This arrangement is ideal, since it ensures that all the animals are continuously exposed to fresh sea water and the current carries faecal and other waste material away. After shell repair has been completed — only 2 or 3 weeks in summer — the spat are ready to be transferred to trays on our oyster ground. This part of the work is described in Chapter 7.

In recent years we have concentrated on the production of cultchless spat and these require careful handling if the best results are to be obtained. An excellent method of culture for the first few weeks after metamorphosis is to transfer the detached spat to shallow trays fitted with a nylon mesh base. These hang at the surface of a tank so that the spat are covered with about 9cm of water. A small plastic pump continuously lifts water from the bottom and sprays it back onto the surface. Since the edges of the trays are above the water surface, the circulated water has to pass downwards past the animals and through the base. If large numbers are involved the trays can be kept in shallow troughs, similar to the flow trays outdoors (plates 14 and 15) and in this case connected to a closed re-circulation system.

The conditions to give the maximum growth have been studied in the laboratory in a series of 50 litre plastic tanks fitted out in the manner described. The tanks were filled with coarsely strained sea water and kept at $21°C$. The plastic trays measured $17.5 \times 11.5cm$ ($201cm^2$) and were stocked with various numbers of spat; the number of spat per litre of water was also varied. *Tetraselmis suecica* was used as food. Many of the experiments lasted for three weeks and food was usually added at a greater rate towards the end of the experiment as the spat grew bigger.

In the first week after settlement the conditions are less critical than later as the spat are very small and only growing slowly. 5000 per tray (2500 per $100cm^2$) through which the water is re-circulated at 1 litre per minute (0.5 litre per $100cm^2$) and feeding 10 cells of *Tetraselmis* per μl is satisfactory.

In the experiments covering the second to the fourth week the amount of food added was found to have a greater influence on growth than most of the other factors tested.

The importance of food is demonstrated in Figure 27a which shows the relation between the size of the spat after 21 days and the total number of cells fed to them in that period. On the other hand, although feeding more cells increases the

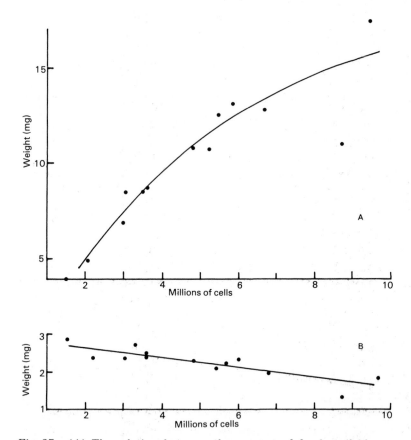

Fig. 27 (A) The relation between the amount of food available per spat, and the live weight of spat after a period of 3 weeks' growth, (B) The weight of spat obtained in 3 weeks' growth for each million cells of *Tetraselmis* fed related to the total number of cells fed in that period.

growth rate the weight of oyster spat obtained for each million *Tetraselmis* cells fed steadily declines — see Figure 27b. The number of spat per litre was also important but almost entirely through its influence on the abundance of food. The number of spat per tray had a small but still appreciable effect. For example, in one experiment where two food levels and three densities per litre were tested, the average weight of native oyster spat after 21 days was as follows:

1,480	spat per tray	8.3mg
877	spat per tray	9.5mg
418	spat per tray	11.0mg

Temperature has an important effect and since we are concerned with a re-circulation system at this stage additional heating need not be an expensive item. The results of one experiment (Figure 28) illustrates the importance of providing an adequate food supply if the temperature is raised. In this trial there was little difference in growth between 14°C and 24°C when the spat were only fed 5 cells *Tetraselmis* per μl per day. If the ration was raised to 10 cells per day some advantage could be taken of an elevated

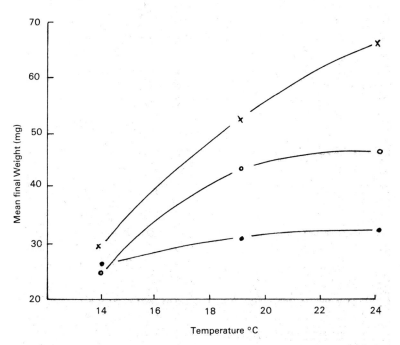

Fig. 28   The relation between temperature and the live weight of the spat after a period of 3 weeks' growth when fed at 3 levels of *Tetraselmis:* ● — ●, fed 5 cells/μ; o — o, fed 10 cells/μl; x — x, fed 20 cells/μl. (From Walne and Spencer, 1974.)

temperature but it was not until 20 cells per day were fed that full advantage was taken of the highest temperature.

From the results of a number of experiments of this type we have concluded that the most rapid growth is given by keeping spat at not more than 250 per 100cm^2 of tray and not more than 50 per litre of water. At about 20°C the feeding regime should be 10 cells per μl per day in the first week, 10 cells in the morning and 5 in the afternoon in the second week, and 10 cells per μl twice a day in the third week. The tanks and trays should be thoroughly cleansed and the water renewed twice a week. In these conditions the spat should reach an average size of 4-5mm within 4 weeks of metamorphosis when they can then be transferred to open flow trays. If less favourable conditions are maintained then the growth rate will be reduced.

The spat are usually moved out of the laboratory environment when they have reached about 10mg in live weight. The general method of culture remains similar as they are transferred to wooden trays (45 × 23cm) with a mesh base. Ten of the trays are housed in a wooden or glass fibre trough (300 × 61 × 15cm deep). Each tray receives a single jet of water at 1 litre per minute from a manifold running along the side of the trough. Water is supplied from a 90 000 litre concrete reservoir which is replenished several times each week. In addition we use a similar system in which the water is recirculated through a 90 000 litre reservoir. This recirculation system is useful in periods when the water is heavily laden with silt or contains a bloom of *Phaeocystis*. The trays are kept covered with sheets of opaque plastic to stop the growth of weed.

Such trays can conveniently hold 2000 spat (2 per cm^2) of 10-20mg, but when a weight of 30-40mg is reached the number should be reduced to 1 per cm^2 and at 50-60mg the density should not be more than 0.5 per cm^2 or growth will be reduced. Servicing comprises regular checks that the water flow is in order and occasional shaking or hosing to remove accumulated silt.

The growth of native oyster spat held in trays of this type on open circulation at Conwy in the summer of 1971 is shown in Figure 29a. This demonstrates that spat reached about 0.1g 4 weeks after leaving the heated recirculation tanks in the laboratory. When planted not later than the end of July, either in trays at Tal-y-foel or on a raft at Menai

138

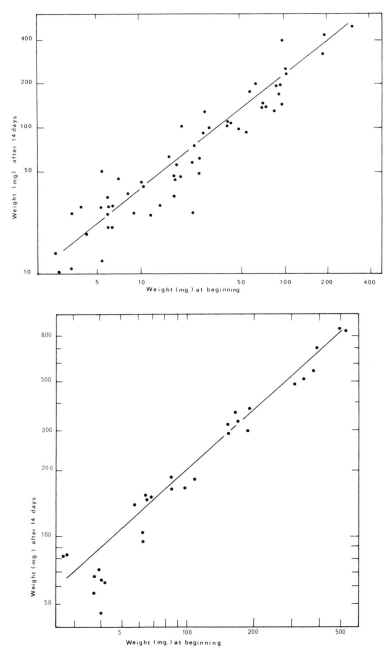

Fig. 29 (A) Growth in 14-day periods of native oyster spat kept in open flow trays at Conwy, (B) Growth in 14-day periods of native oyster spat kept in trays on a raft and on the shore in the Menai Straits.

Bridge, these spat will reach 1g in weight by the winter (Figure 29b): they increased their weight tenfold in 2 months.

SELECTED REFERENCES

SHELDON, R W and T R PARSONS, 1967. A practical manual on the use of the Coulter Counter in marine research. Coulter Electronics Sales Company, Canada. 66 pp.

WALNE, P R, 1970. Studies on food value of nineteen genera of algae to juvenile bivalves of the genera *Ostrea, Crassostrea, Mercenaria* and *Mytilus. Fishery Invest., Lond.*, Ser. 2 **26 (5),** 62 pp.

WALNE, P R and B E SPENCER, 1974. Experiments on the growth and conversion efficiency of the spat of *Ostrea edulis* L. in a recirculation system. *J. Cons. int. Explor. Mer.,*35(3): 303-318.

6 Other species of bivalves

For many years the rearing work at Conwy concentrated on the European flat oyster, but once a technique had been developed the next logical steps included applying the methods to other species. Hatchery culture is essentially the domestication of a marine species, and it should not be supposed that the species which are native to a country necessarily include all those which are most suited to culture. Many of the animals and plants now used in agriculture and forestry were not developed from native species, and it is very probable that as marine culture expands we shall see many strange animals growing in British waters.

As with many developments, there are potential dangers in the introduction of a new species. Oyster farmers in several areas of the world have suffered to a very substantial degree from pests introduced by the transplantation of oyster stocks from one area to another. The introduction of the American slipper limpet (*Crepidula fornicata*) and the American tingle (*Urosalpinx cinerea*) on to the south-east coast of England from the Atlantic seaboard of North America are well-documented examples. A measure of protection against a repetition in the British Isles is now provided by the Deposition of Shellfish Order (1965, 1966). If foreign shellfish are introduced through a hatchery, only small numbers of stock for breeding purposes need be imported and these can be kept in strict isolation. The young which are bred from these will be free from pests.

The potential danger which is more difficult to predict is the likely ecological effect of a new species when it is released in a new area. It is not uncommon for an exotic species, freed from the natural checks which operated in its original habitat, to proliferate to an exaggerated degree. The example of the rabbit in Australia will be familiar to many readers. If the introduced species competes with established forms of economic importance, this could have significant consequences which may not necessarily be harmful. The total value of the production from a given area might well go up. If, however, the introduced species proliferates in an area which does not already support an important commercial crop, the advantages could be considerable, although undoubtedly the ecology of the area would be changed.

The dangers can therefore be minimized either by selecting species which will occupy areas not inhabited by other commercial species or, alternatively, by choosing forms which are unlikely to breed in British waters. The quahog, which is discussed at length later in this chapter, is an example of the former. The Portuguese oyster provides a good example of the latter. Stocks of this species of oyster have been imported from Portugal for many years and relaid on the south and east coasts of England where it grows and fattens extremely well, yet only in the most favourable situations does it breed and then only to produce a meagre spatfall. Thus, although it is cultivated adjacent to beds of flat oysters, it causes no harm; by contrast, in the Bassin d'Arcachon, where it reproduces freely, its introduction has led eventually to it supplanting the native species.

The remainder of this chapter gives details of our experience in rearing six species of bivalves, five of which are not native to the British Isles. These species were selected for a variety of biological reasons, because they could be obtained with reasonable ease, and finally because it was judged that they were likely to be acceptable to Western European palates. For many species of bivalves their biological requirements are insufficiently known for it to be possible to judge how well they will survive and grow in a new area; by far the most satisfactory way is to breed a small stock and test these.

Ostrea chilensis, THE CHILEAN OYSTER

This is a small species of oyster with a distribution centred on the indented and sheltered coast of southern Chile. Little was known in the scientific sense about this species, and interest was aroused when Dr R H Millar of the Millport laboratory drew attention to a report made by the University of Lund after its 1948 expedition to Chile. This stated that the scientists had collected oysters containing eggs and larvae which measured 0.5-0.75 and 0.7-1.2mm respectively. These are very large for any bivalve and it seemed likely, on theoretical grounds, that the larvae would not be planktonic for more than a few hours; if so, then they would avoid many of the hazards to which the planktonic stage is liable. Moreover, from the commercial viewpoint, the larvae would set close to their parent stock and not be scattered over a wide area.

The major fishery for this species is on the island of Chiloe. Although this is at a latitude equivalent to northern Spain, the climatic regime is more like that of England because of the cooling influence of the Humboldt current. The winters are, however, mild and this proved to be an important point.

A sample of these oysters was flown to Conwy in September 1962 and numerous mature larvae were liberated in October and November. The average shell length of various batches varied between 435 and 492μm, with individual larvae ranging from 390 to 520μm. As expected, most of these settled within a few hours on the glass side of the tanks, on wood and on pieces of shell. Some larvae which were isolated in a clean glass beaker were still swimming 24 hours later, but were able to attach to a piece of shell. Presumably because of the size of the larvae, each oyster produced only about a tenth of the number which would be produced by the equivalent sized European oyster.

The spat were very small by the time the winter came, so they were kept indoors until the spring. It was difficult to provide sufficient food and water during what turned out to be the exceptionally severe winter of 1962-63, and by the spring the survivors were somewhat stunted. However, they reached 2-3cm in diameter in the autumn, when they were planted on our Tal-y-foel oyster ground; none survived the winter out of doors, and tests showed that they could not withstand quite moderate frost.

From these trials, and other information obtained during a visit to the Chilean grounds, it has been concluded that this oyster is unlikely to be generally useful in Britain. It will not withstand the winters encountered on most, if not all our grounds, and it is also rather small. Rarely does it grow to more than 6-7cm in diameter. It is possible that we shall be able to exploit the biological advantages of a reduced planktonic phase by using the New Zealand oyster, *Ostrea lutaria*, which also has this peculiarity.

Ostrea lutaria, THE NEW ZEALAND OYSTER

This species of oyster is native to New Zealand where it supports a considerable fishery, especially in the Fouveaux Strait at the southern end of the South Island. Like *O. chilensis* it has an abbreviated planktonic stage (Hollis and Millar 1963), but the adults grow to a larger size. The minimum legal size is a diameter of 5½cm but those landed are usually 6-9cm. Oysters brooding larvae are found from August to March but the majority of the larvae are released in December and January. Many of the populations are found on shingle or shell bottoms in areas of clear water and strong tides, and their shells, which are similar to those of *O. edulis* in appearance, have the flat shape characteristic of oysters grown on a hard bottom.

The first batch of *O. lutaria* arrived in Conwy in the late summer of 1963 and numerous liberations of larvae were obtained soon afterwards. We were then faced with the problem common to all animals imported from the southern hemisphere; the period when they are ripe and ready for breeding is our autumn and winter. It is most satisfactory to obtain the breeding stock in a ripe condition since this reduces the period for which they have to be kept, and therefore diminishes the risk of pests getting free. We have found it difficult to keep breeding stock in good condition for long periods when held in isolation; the quantities of food and water required are very great and it is usually much more practical to obtain ripe animals. When breeding has been accomplished then the very small spat have to be kept until the water is warm enough for them to start growing in the following spring.

The larvae settled as quickly and readily as those from the Chilean oyster. Survival of the smallest spat kept outdoors during the winter was good — this species is apparently unaffected by the normal winter conditions in North Wales — but growth during the following spring and summer was very poor. The survivors could not be stripped off the standard plastic collectors until the late summer, and in the autumn were planted in a tray at the oysterage. These survived the winter well, but unfortunately the tray was washed away in the spring of 1965. Fortunately, however, a sample of nine was being kept in another tray with the same number of *O. edulis* for comparison. The average weight of the oysters in these two groups has been:

| | Mean weight (g) | |
	O. edulis	O. lutaria
April 1965	0.84	0.95
October 1966	26.5	22.2
September 1967	43.2	39.5
October 1968	62.0	51.0

No deaths occurred until the final year and this small sample suggested that it should be possible to find areas in British waters in which the New Zealand oyster will grow and survive reasonably well.

A further stock was imported in July 1966 and by 15 October liberations of larvae had occurred. In an attempt to avoid stunting, some of these spat were kept in warm water. Although they could be stripped from the collectors during the winter, few managed to heal the shell damage and survive. The rest, which remained outdoors still attached to their collectors for the winter, were removed in the following June. Mortality was again heavy; 70 per cent died in the next 6 weeks before being transferred to the Tal-y-foel oyster ground, where 90 per cent died in the following 7 months. A handful survived in a tray and in November 1970 the shells of these were observed to carry a number of New Zealand oyster spat a few millimetres in size, which must have come from the successful reproduction of these four-year-old oysters in the summer of 1970. A careful search failed to find any spat attached to the wire or frame of the tray.

Up to the present, then, our experience with this species has been disappointing. Although the stock has been imported successfully and has bred readily, the survival and vigour of the young spat obtained has been poor. It may be that this species requires certain conditions in its environment which we have been unable to provide.

Crassostrea gigas, THE PACIFIC OYSTER

Species of *Crassostrea* support the major oyster fisheries of the world. None is native to the British Isles, although *C. angulata*, the Portuguese oyster, is regularly imported for relaying. The Pacific oyster is abundant in Japan and seed are regularly imported from there into the State of Washington in the United States, and into British Columbia. Data given by Quayle (1950) showed that this oyster had a very high growth rate at quite low water temperatures (15-19°C), and this suggested that it would be a worthwhile species to test in Britain.

A small stock was obtained from Pendrell Sound, British Columbia, in June 1964, and releases of eggs and sperm were obtained in July and August. The stocks of this species in commercial use in Britain in 1973 are the descendants of this importation. In order to increase the size of the gene pool a further importation was made in 1972 from a very isolated population in Seymour inlet on the west coast of Vancouver Island. The eggs of this species are released into the water for external fertilization and, when the parents are ripe, this can be stimulated by placing the animals for an hour or so in water of 27-29°C. The addition of some sperm suspension obtained from the gonad of an opened oyster may also be required. Trials with Conwy-reared stock showed that from oysters brought into the hatchery in January and kept in running water at 21°C spawning could be obtained after about 2 months, although eggs and sperm stripped from opened oysters after 5 weeks' conditioning could be success-fully reared. In the summer and autumn, spawning could be stimulated within a week of bringing stock into the hatchery.

The method of treating the fertilized eggs is exactly the same as that described below for quahogs. After 24 hours at 23°C the larvae are fully shelled but they are very small, only $35\mu m$ in length. In the first experiments the growth rate of these small larvae was very slow and survival was poor. Various combinations of food supply, temperature and salinity were tested and eventually it was found that satisfactory results could be obtained by rearing the larvae in normal Conwy water (salinity 30-32 per thousand) at 28°C. The water was filtered and food provided in the form of a mixture of small algae, particularly *Isochrysis galbana*, *Cyclotella nana* and *Chaetoceros calcitrans* in equal numbers to give a total concentration of 100 cells per μl. When the larvae reach $180\mu m$ in size the food mixture which we use for the flat oyster, 50 cells of *Isochrysis* and 5 cells of *Tetraselmis suecica* per μl, is suitable. With this method we were able to rear 9 per cent of 1.75 million larvae through to spat in a culture made in 1970.

The spat have shown good survival and excellent growth. In trays at Tal-y-foel the average monthly mortality in the summer is about 2 per cent, and spat over about 5g in weight increase in weight by about 5g per month during the months of May to September. In 1967 the growth was compared in different parts of the country. In the spring samples of 50 oysters with an average weight of 15.5g were planted in trays; these were next examined in the autumn and the average weight found is shown in Table 7. Growth was very good; in the best places (River Yealm, River Crouch and Newtown river, Isle of Wight) the total weight increased five to seven times. Compared with our native oyster, this species starts growing earlier in the season (see Figure 36) and is therefore more able to take advantage of those times of the year when food is abundant. It should do well in a wide range of grounds in Britain and will be a useful species for hatchery rearing. It is probable that, like the Portuguese oyster, it will spawn only in very favourable areas, because the water temperatures will generally be too low for the stimulation of the release of the sexual products. Samples of *C. gigas* with an average size of about 12cm have been sent to a number of growers accustomed to handling Portuguese oysters. The comments on shape were most unfavourable, but this was doubtless due to the oysters having been reared in trays. The appearance of the shell generally attracted favourable comment and the flavour of the oysters was appreciated. Several planters laid them on the shore or in pits and were surprised that they withstood frost so well. Their growth in the lower part of the inter-tidal zone is also good. In 1971 trials made at Tal-y-foel compared the growth in trays at levels corresponding to 2, 4, 6, 9 and 15% exposure. In the period March to December Pacific oysters with an initial live weight of 4g grew to 58-67g at up to 9% exposure. At 15% exposure growth was reduced to 47g.

The shape of *Crassostrea* oysters is often strongly influenced by their environment, and those in these samples, which were cage-grown, were markedly distorted. Planted on suitable ground, no difficulty should be experienced in producing the correct shape, but care will have to be exercised in marketing them at the right size. If left, they quickly grow too big. For example, at Tal-y-foel they have reached over 25cm in length in 4 years (Plate 24).

Venerupis decussata, THE PALOURDE OR BUTTERFISH
This species is known as *la palourde* in France, where it is very highly esteemed. The French name can be conveniently

adopted in Britain to avoid confusion with other species that are also called butterfish. The palourde is not uncommon in Britain, buried deeply in muddy grounds, but so far has not been found in commercial quantities.

A few were obtained from Southampton Water in the summer of 1966 and it was not difficult to induce them to spawn; two batches of larvae, 916 000 in all, were reared and grew well, yielding 540 000 spat. These seemed very hardy when kept in the flow trays under the conditions which we normally use for *Mercenaria* spat, but their growth rates were comparatively slow. Our recent experience suggests that this species does much better when it is allowed to burrow. Several samples of spat have been planted at Tal-y-foel, in Poole Harbour and in the River Yealm, and the annual increase in shell length found is illustrated in Figure 30. These animals had been allowed to bury in the mud at the low-water mark of spring tides and were guarded by a plastic mesh as a protection against predation. This technique is described in more detail in the section on quahogs. The

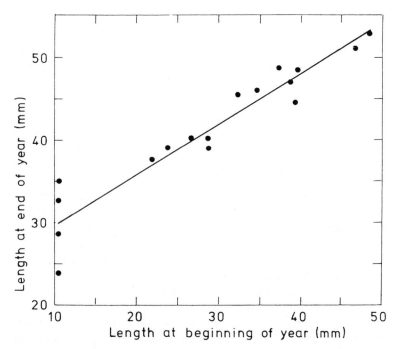

Fig. 30 The annual increase in size by the palourde (*Venerupis decussata*).

annual growth increments observed suggest that spat planted at 10mm in size should exceed 45mm after 3 years; this rate of growth is similar to that found in the quahog (see p. 161) although growth in the first year is more rapid.

A major difference between this species and the quahog is that the quahog buries to just below the surface of the soil, whilst the palourde buries to a depth of 7-10cm and will therefore be more difficult to harvest.

Choromytilus choros, THE CHILEAN MUSSEL
This species occurs among the islands in the southern part of Chile, where it is a valuable, though scarce, crop. It is similar to the common mussel in external appearance, but the flesh is more highly coloured. In Chile the minimum legal size is 10cm (4 inches) in length, individuals of 20cm (8 inches) are not uncommon, and it may grow to 30cm (12 inches). These sizes, and the appearance of the shell, suggested that it might have a more rapid growth rate than our own mussel.

A consignment was imported in the autumn of 1965, but as these had already spawned in Chile we were unable to stimulate them to release the sexual products until the following spring. A satisfactory stimulation was given by removing the animals from the water and injecting a few ml of a half-molar solution of potassium chloride into the mantle cavity. The animals were replaced in the water after about half an hour, and the release of eggs and sperm by ripe individuals would commence shortly afterwards. This procedure, first introduced by Iwata (1951) for the common mussel, causes little mortality and we have found that the same animal can be repeatedly injected over a period of several weeks.

Fertilization was readily obtained and swimming blastulae were present in good numbers after 24-28 hours, but further development was very poor and slow. We repeated the experiments numerous times but we were fortunate if half a dozen spat were obtained from half a million larvae. Eventually the trouble was traced to the temperature at which the larvae were being reared. All these batches had been cultured at the same temperature as that used for oyster larvae, ie. 21-22°C. That this is too high for *Choromytilus* was shown by rearing a batch of larvae with a mean length of 110μm for 6 days at a variety of temperatures, with the following results:

Mean temperature (°C)	18.5	18.7	21.0	22.5
Final size (μm)	161.3	163.1	149.3	131.8

Poor results are also obtained if spat are cultured at temperatures above 20°C. For example, in one experiment, groups of 10 spat with mean lengths of 3.4-4.0mm were cultured at several temperatures. After 14 days the lengths were:

Mean temperature (°C)	15.3	18.1	20.4	22.1	23.8
Final size (mm)	6.7	6.4	6.2	4.7	4.2

This was our first experience of a species which had a marked upper temperature limit, but once this was understood some good batches of spat were reared.

Samples of spat were tested in various parts of the country in 1967, and the results (Table 7) showed a reasonable growth rate which was not markedly different from that found in the native mussels. The range of weights between different places was rather high, which suggests that this species is fairly sensitive to small differences in the environment. At present we do not see that the Chilean mussel offers any special advantages over the native one.

Mercenaria (Venus) mercenaria, THE QUAHOG

This species of clam is found along the east coast of North America, from Prince Edward Island and Nova Scotia in the north to Florida in the south. The major populations, which support flourishing fisheries, are centred in New England. According to the US Fishery Statistics the hard clam crop in 1965 was about 50 000 tons, which had a value at first sale of about 10 million dollars. Quahogs are typically sorted into three basic grades, as follows:

Shell length (inches)	*Grade*
2-2½	Little neck
2½-3	Cherrystone
3+	Chowder

In some fisheries further subdivisions of these may be made. The smaller grades are usually eaten either freshly opened or steamed, whilst 'chowders' are processed into a soup which is often canned. The value of a quahog is approximately constant throughout life once it reaches a commercial grade — about 2 cents in 1967. The value per unit weight therefore steadily falls; in 1966 a bushel of little necks was worth about $15, of cherrystones $10, and of chowders about $4.50.

For clarity it is best to use the old Indian name of 'quahog' for this species, because it avoids confusion with other species. For example, the scallop is called a clam in Scotland and the palourde, dealt with earlier in this chapter, is also known as a clam.

The larger populations of quahogs, which are below low-water mark, are found on soils varying from soft mud to shell and mud. They can be readily fished from soft mud by a hydraulic dredge which cuts a trench a few inches deep through the bottom. The mud is blown clear by jets of water from a manifold fixed in front of and above the cutting edge. The quahogs are then brought to the surface either by hauling the dredge up or by arranging for a conveyor to bring them up continuously. Such a procedure can be very efficient; boats in Great South Bay on Long Island typically catch 15 to 20 bushels a day.

From this account it is clear that this is an important commercial species in North America, where a considerable technology has been developed for the mechanization of fishing, grading and processing. Its biology has also received particular attention at the Milford laboratory of the US Bureau of Fisheries, where it was shown that the larval culture of quahogs was usually easier than for oysters. Heppel (1961) has reviewed the numerous attempts to establish this species in Britain, France, Holland and Belgium; most of these were unsuccessful, although some reproduction has been recorded in Brittany (Marteil 1956).

A small sample of seed quahogs which had been reared at the Milford laboratory was received at the Ministry's Conwy and Burnham-on-Crouch laboratories. Most of those that were laid out in the Essex rivers suffered severe predation by crabs, whilst those at Conwy grew very slowly. This was discouraging and the matter might have languished for some time if it had not been for the discovery of a substantial population in Southampton Water (Ansell 1963). Although the quahogs were limited to the upper part of the estaury, some beaches, for example those adjacent to the Marchwood power station, carried 50 to 150 per m² in a broad band from mid-tide level to the low-water mark of spring tides. The origin of this population is unknown; growth rings on the shell show that there were heavy spatfalls in 1959-61, but considerably older individuals have been found.

The convenience of this population as a source of breeding stock, and its demonstration that quahogs could thrive in Britain in a highly polluted and apparently not very promising area, stimulated comprehensive studies based on Conwy-reared spat.

The general procedures for culturing the breeding stock and larvae have been similar to those used for rearing *Ostrea edulis*. The main differences are, first, that it is practicable to stimulate the spawning of ripe adults by raising their water temperature several degrees above the conditioning temperature for a brief period, and, secondly, that it is unnecessary to use spat collectors, because the larvae do not become permanently attached at the end of the free-swimming period.

It has become our practice to rear quahog larvae in the late winter and early spring, in order that the spat may gain the maximum advantage of the summer growing period before reaching their first winter. We have found that spat reared in July or August will reach only 1-2mm by the end of the autumn, and although many will survive the winter, there is a pronounced tendency for the renewal of growth to be delayed until well into the following summer. Even then, the spat may remain permanently stunted.

Quahogs measuring 4-7cm are brought to Conwy from Southampton in December and January and kept in warmed (20-22°C) running sea water in the tank room. During this period the gonads gradually develop, and by making periodic attempts to induce spawning, it becomes apparent when ripe individuals are present. The following table, which indicates the length of the conditioning period for Southampton quahogs for three successive winters, shows how long it may be necessary to wait:

Winter	Conditioning started	Number of days to spawning
1964/65	December	29
1965/66	January	52
1966/67	December and February	37

Spawning is usually accomplished in a polythene bowl holding about a dozen quahogs. A vigorous flow of sea water heated to about 26°C passes through the bowl and, when the siphons are well extended, a few ml of a suspension of male gonad prepared by chopping pieces of tissue in water are added. Once an individual quahog, either male or female, has started to spawn, it can be taken out of the bowl and placed in another vessel filled with clean water. It will very soon start spawning again and in this way eggs and sperm can be obtained relatively free from debris.

The eggs are fertilized in 1 litre glass beakers by adding a few ml of sperm suspension. The eggs slowly sink to the bottom and surplus sperm are removed after about half an hour by decanting the supernatant water and replacing with freshly filtered water. To avoid crowding the developing eggs, it is best to transfer them to a suitable glass tank so that they are spread as a thin layer on the bottom. After 24 hours at 22°C the embryos are swimming and on the second or third day the shell is sufficiently developed for it to protect completely the larva. Once this stage has been reached, the larvae can be collected on a sieve and the water changed in the usual manner.

We usually rear the quahog larvae in 75 litre polythene bins by the same procedure as that described in Chapter 3 for oysters. Since the early larvae are much smaller than those of *Ostrea edulis* — about 90 microns compared with 170 microns — we can raise them at much higher densities. Up to half a million in 75 litres gives as good a result as smaller numbers. In three years we reared 18 batches of larvae, with the following results:

Year	Number of batches	Average initial number of larvae	Percentage metamorphosed	Days to metamorphosis
1964	8	268 000	50	20
1965	7	642 000	43	23
1966	3	399 000	33	19
Average	—	436 000	42	21

These suggest that an average production of about 183 000 quahog spat can be expected from a 75 litre culture after a planktonic stage of 21 days. The larvae can also be readily cultured in the 12 litre glass tubes described on p. 72; in these, normal growth is obtained at densities of 40 000 to 60 000 per litre, provided that additional food culture is added continuously with a dosing pump.

Metamorphosis of the larvae occurs at a shell length of about 250 microns. At this stage the ciliated foot becomes prominent and the larvae start to crawl on the bottom. This is accompanied by the production of a byssus thread with which the larvae attach themselves to the sides or bottom of the culture vessel. This alteration in behaviour is clearly seen when the water in the bin is changed, since many of the larvae remain behind when the water is removed, and can only be washed from the walls with a fairly strong jet of water. A few larvae may reach this stage after 16 days and usually 75 per cent have metamorphosed about 21 days after fertilization; by then the production of byssus is so plentiful that the spat are coalescing into clumps of several thousand individuals, and a stagnant culture becomes unsuitable.

To overcome the smothering effect of these clumps the spat are transferred to trays, with a nylon mesh base, immersed in sea water. By circulating sea water continuously through these trays even the spat in the middle of the bunch stand a chance of receiving some of the water. We use trays which are 8cm deep and with a floor area of about 400cm^2. The frame is made of wood and painted with several coats of a polyurethane varnish; this forms a hard shiny surface which can be readily cleaned. The mesh base is stapled to the bottom of the frame and sealed into place with varnish. Eight trays stand in a large wooden holding tray and water runs continuously from a high-level tank into each of them and then into the holding tray, from which it overflows to a tank on the floor. The high-level tank has a heater and thermostat to control the water temperature at about 22°C. A pump periodically pumps the water back from the lower tank up to the holding tank. This closed circulation system holds a total volume of 450 litres, and good growth and survival can be obtained with up to 750 000-1 000 000 quahogs at a time. Originally we used filtered sea water but coarsely strained water enriched with a mixture of 50 cells of *Isochrysis* and 5 cells of *Tetraselmis* gives better results. If a full stock of

154

quahogs is being carried it is necessary to add algal culture continuously.

The water in the system is renewed twice a week, and at this time the spat are thoroughly washed to remove detritus and to break up the clumps caused by byssus formation. As they grow the spat are moved on to progressively larger-meshed trays, which more readily allow the flowing water to remove faeces and other detritus. Mesh sizes which we have found convenient for this and the subsequent stages in quahog rearing are as follows:

Approximate length of quahog (mm)	0.25	0.5	0.8	1.2	2.0	3.0	5.0	10.0
Mesh size (mm)	0.124	0.210	0.440	0.611	0.780	1.550	3.2	6.4

As the spat progress, the range of sizes in a population increases and better results are obtained if the larger animals are separated by selective sieving; this becomes a constant task throughout the time that the spat are kept in trays. As in any natural population there is a considerable range of individual growth rates, and in commercial practice the poor growers would soon be eliminated.

Since unfiltered water is being used the larvae of other marine animals are brought into the system and will rapidly grow in the conditions of gently flowing water and abundant food. However, this can be readily controlled by washing the spat with fresh water to which occasionally some commercial hypochlorite has been added.

When the spat reach about 1mm in shell length they are removed from the closed system and placed in running sea water at ambient temperature outside the laboratory (Plate 16). The main reason for this move is the substantial quantities of food and water which would be required for half a million or more spat over 1mm in size. By this time the unit of 450 litres requires about 5 litres of algal culture per day. The success of this stage is shown by the following table:

Year	Number of batches	From larval vessel		Mean size when moved outside (μm)	In closed system	
		Mean size (μm)	Mean number per batch		Days	% survival
1966	4	272	119 500	1346	44	68.8
1967	3	263	314 000	1093	41	65.0

This result, combined with those for the larval stage already given, shows that over an average period of 63 days we reared about 28 per cent of our fertilized eggs to a mean size of 1.2mm. Since many millions of eggs can be obtained from an average adult quahog, this stage in the life cycle can be very productive.

When moved outdoors the quahogs are again in trays with a mesh base, so that the water continually flows from the surface down past the animals. The arrangement adopted is illustrated in Plate 16. Each tray has an area of 900cm² and as the quahogs grow they are moved on to successively larger plastic meshes; the larger the mesh, the less likely is it to be choked with detritus.

Since we are interested in different batches of quahogs we use rather small trays; eight of them were fitted in a trough 240 × 60cm by 15cm deep. Water was supplied to each batch of quahogs from 1 inch PVC pipes on the two long sides of the trough, and this arrangement conveniently provided a flow of about 1.2 litres per minute to each tray. Since we kept the flow rate constant, the number of quahogs which will grow well in a tray depends on their size; the numbers which we have found satisfactory in our trays are given in Figure 31. At the flow rate of 1.2 litres per minute and stocking densities shown in Figure 29 the quahogs are removing about 75 per cent of the particulate matter in the water during the summer months. The trays require regular maintenance: a daily check to ensure that the water is flowing correctly, a hosing once a week in summer to remove accumulated mud, and the cleaning of the inlet pipes about once a fortnight to remove sedentary organisms such as barnacles and mussels.

It is clear that to keep quahogs under these conditions entails the use of considerable quantities of sea water. From

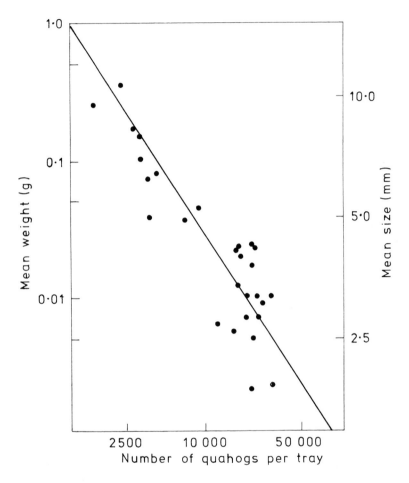

Fig. 31   The relation between quahog size and the number per 900cm² tray (curve drawn by eye).

the figures cited above it can be calculated that a million quahogs require as follows:

Size of quahogs (mm)	2.5	5.0	10.0	15.0
Seawater requirement (litres per hour)	2058	9000	36 000	144 000

Assuming that the water has to be raised 20 feet above sea level, the water requirements of a million 10mm quahogs necessitate a 1 hp pump running continuously, while by 15mm they will require a 4 hp pump.

From this it follows that the siting of any trays requires very careful consideration, and there are substantial economies in getting the young spat planted into the sea as soon as possible. We encountered difficulties in this due to predation by the common shore crab, *Carcinus maenas*, and the following paragraphs show the extent of the problem and how it can be overcome.

The first attempt to plant out quahogs of about 1cm in size were quite unsuccessful. We laid a few hundred at the low-water mark of spring tides and all that remained 2 weeks later were a great number of shell fragments. It was probable that this was due to the green shore crab, *Carcinus*, so we started a series of studies to find the relation between the size of crab and the size of quahogs which it would eat. Armed with this information it is possible to decide between keeping the quahogs in tanks until they are large enough to be immune to crab predation or, alternatively, planting them at a smaller size but protected from crab attack.

Individual crabs of various sizes from 15 to 80mm carapace width were kept in small plastic boxes submerged in a tank of sea water which also circulated through the boxes; five quahogs of each of three sizes were placed as food in each box. The size range was chosen (after some preliminary experiments) so that few, if any, of the larger group were eaten, one or two of the middle group, and perhaps three to four of the smaller group. The number of quahogs which had been eaten was counted each day and the number made up. This was continued until the crab had eaten some quahogs on 7 days.

The results are shown as a series of histograms in Figure 32. Ten crabs of each size group were tested and the

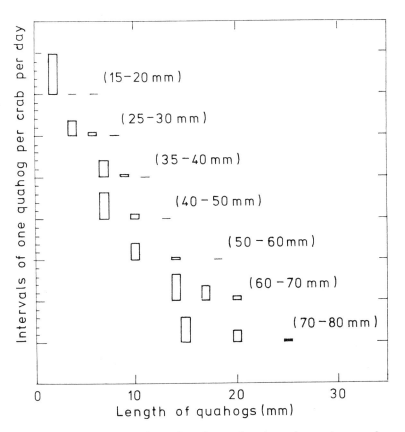

Fig. 32   The mean number of quahogs of various sizes eaten per day by shore-crabs. The figures in parentheses show the carapace width of the crabs.

height of the vertical bars is proportional to the average number of quahogs eaten each day. The nature of the experiments restricted the total number of quahogs that a crab could eat; for example, the 35-40mm crabs ate on average more than three out of five of the 5mm quahogs. The occasional one not eaten was probably because the crab failed to find it. If sufficient quahogs had been provided so as to give the crabs an infinitely large diet, the amount of work required would have been considerably increased. However, other tests showed that such experiments would not have increased the size of quahog that a crab of a given size could tackle — and that was the main point of interest.

The results suggest the following relation between sizes of predator and prey:

> Quahogs of 7mm are not attacked by crabs of below 20mm carapace width
>
> Quahogs of 14mm are not attacked by crabs of below 50mm carapace width
>
> Quahogs of 20mm are not attacked by crabs of below 60mm carapace width
>
> Quahogs of 25mm are not attacked by crabs of below 80mm carapace width

This table gave us a guide to the degree of protection required if quahog seed was to be successfully planted on the shore. Some tests showed that this could be achieved if a mesh of suitable size was laid on the surface of the mud in which the quahogs were buried. This excluded the crabs and helped to retain the quahogs when they were first laid. Good survival of groups of 200 quahogs could be obtained by confining the group in a wooden frame 60cm square and 8cm deep, which was sunk flush with the surface of the beach. The top of the frame was closed by a plastic mesh; 6, 12 and 18mm were the sizes used. The cover extended about 15cm beyond the frame and was fixed down to the surface of the mud with wire pegs (Plate 17). Little obstruction was offered to the flow of water by this equipment, which was also unobtrusive to the casual observer. We found that two types of mishap could occur; the mesh could be blocked by a growth of weed and Ascidians, or the frame could be covered with a layer of silt. Since the mesh restricted the movement of the quahogs, either of these factors would reduce or even stop access of water to them. However, these troubles occurred only occasionally and the method allowed us to test the growth and survival of seed quahogs in a considerable variety of habitats on the coast of England and Wales.

The quahogs used in these trials were rigorously graded by selective sieving, so that the range of sizes present at a laying was quite small. Samples with average sizes varying between 9 and 21mm were laid at nine general areas on the south-east, south and south-west coast of England, and at the Tal-y-foel oysterage, in 1966-68. The quahogs were visited in the spring and autumn, and their average size, and hence their growth rate, was determined. In addition, in some cases the total number of live quahogs was found by sieving the top few inches of soil.

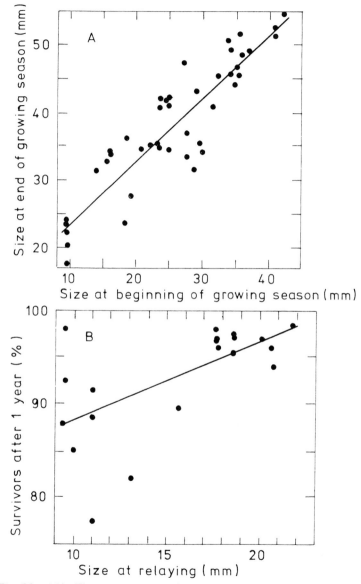

Fig. 33 (A) The mean size of relaid quahog spat in the spring (horizontal axis) related to their size in the autumn (vertical axis). (B) Mortality in first year.

The annual growth rate found in these trials is summarized in Figure 33a. This diagram relates the mean size of the quahogs, in each frame, at the beginning of the growing season to the size that they reached by the autumn. The line represents the average growth obtained in all the results and can be used to calculate the average size reached in successive years by relaid quahogs. It should be remembered that the trial sites were selected to represent a variety of areas, whereas in commercial practice the more favourable grounds would be selected, and the growth rates could be higher than the average values shown in Figure 33a.

The quahog has a restricted growing season; in Conwy this extends from May to September, and it is sharply limited to this period by the water temperature. Observations of the growth of small quahogs kept in trays in a number of places in 1967 show that the best results are obtained in the warmer areas — the south and south-east coasts of England (see Table 7).

Twenty plots which were exposed for periods varying between 14 and 34 months gave reliable information from which annual survival can be calculated. This, expressed on a percentage basis, and related to the size of the quahogs at the time of planting, is shown in Figure 33b. The points appear to fall into two clear groups: one relates to animals which were 9-13mm in size at the time of planting and which showed a survival rate of 88 per cent per year, and the other relates to animals which measured 17-21mm at planting and had a survival of 96-97 per cent per year. It is not surprising that the smaller animals had a higher mortality.

The information contained in Figure 33 can be used to forecast the likely average outcome of quahog relaying. Taking 10mm seed at the time of planting and using a survival figure of 87.9 per cent for the first year and the higher figure of 96.7 per cent for subsequent years, the average results are:

	Increase in length during year (mm)	% of the original planting surviving at the end of the year
First year	10.0-23.2	87.9
Second year	23.2-35.6	85.0
Third year	35.6-47.2	82.2
Fourth year	47.2-58.0	79.5

In these trials the quahogs were laid at a density of 2 million per acre. It is unrealistic to suppose that in large-scale working 80 per cent will be recovered after 4 years, but it is not improbable that 50 per cent could be. The weight of the million quahogs, 55-60mm in size, would be 30 metric tons.

The quahog is excellent material for culture in a hatchery, and there is no doubt that it will grow and survive well in many places in the British Isles. It thrives in soft muddy areas which are quite unsuitable for other types of shellfish culture, and trials by the White Fish Authority have shown that it can be readily harvested from this type of soil by a hydraulic dredge. However, a large-scale method for planting the spat is still required. Clearly there are formidable problems involved in covering extensive areas with netting. Similarly control of such an ubiquotous pest as the green crab will not be easy.

SELECTED REFERENCES

ANSELL, A D, 1963. *Venus mercenaria* (L.) in Southampton Water. *Ecology*, 44, 396-397.

HEPPEL, D, 1961. The naturalization in Europe of the quahog, *Mercenaria mercenaria* (L.). *J. Conch., Lond.*, 25, 21-34.

HOLLIS, P J and R H MILLAR, 1963. Abbreviated pelagic life of Chilean and New Zealand oysters. *Nature, Lond.*, 197, 512-513.

IWATA, K S, 1951. Spawning of *Mytilus edulis*. (4). Discharge by KCP injection. *Bull. Jap. Soc. scient. Fish.*, 16, 393-394. (In Japanese, English summary.)

MARTEIL, L, 1956. Acclimatation du clam (*Venus mercenaria* L.) en Bretagne. *Rev. Trav. Inst. (scient. tech.) Pêch. marit.*, 20, 157-160.

QUAYLE, D B, 1951. The seasonal growth of the Pacific Oyster (*Ostrea gigas*) in Ladysmith Harbour. *Rep. prov. Dep. Fish. Br. Columb.*, (1950), 85-90.

WALNE, P R, 1963. Breeding of the Chilean oyster (*Ostrea chilensis* Philippi) in the laboratory. *Nature, Lond.*, 197, 676.

7 The Tal-y-foel oysterage

As soon as the oyster spat, or indeed any of the species which have been reared at Conwy, were sufficiently large and robust to be planted in the sea they were removed to the Tal-y-foel grounds in the Menai Straits. These are 26 miles from Conwy and are the nearest sheltered and accessible area for this type of work. This distance has been an obstacle to detailed observation and in an ideal situation the planting grounds should be adjacent to the hatchery. In any new development care should be taken to ensure that sheltered grounds, suitable for holding a wide variety of bivalves, are close at hand.

The Tal-y-foel oysterage is at the south-western end of the Menai Straits on the Anglesey shore. The lower part of the beach, which is exposed only on spring tides and is the part used by us, is protected from the effects of rough weather by a sandbank, Traeth Gwyllt, which separates the Tal-y-foel inlet from the main channel. A tidal current of 1-2 knots flows parallel with the shore, associated with a tidal range of about 17 feet at springs and 10 feet at neaps. Low water of spring tides is in the late afternoon or early morning. There is steeply sloping shingle at the top of the beach; below this the slope is much more gradual and the soil varies from mud, with varying amounts of sand, to small shingle bound together by mud and sand. Further up the inlet there is a small exploited mussel bed, and the whole area is occasionally picked over for periwinkles. At one time the grounds were used for relaying oysters imported into Liverpool from North America, and *Crassostrea virginica* shells can occasionally be found.

The soft ground makes it difficult to move equipment across the beach, but much of this difficulty has been overcome by laying a T-shaped track with the spine running down the beach and the horizontal arm at low water. A few inches of 5cm roadstone laid on moderately soft mud will consolidate into a hard track. At first we used a hand trolley, but the path is now sufficiently strong for a tractor or a motor vehicle with four-wheel drive. A further dressing of stone every few years and attention to any ruts which may form is all that is required for maintenance.

There are no major streams in the area and a series of samples taken at low-water springs over a 2-year period shows that salinities lower than 31% were rare; usually they were in the range 31-33%. The mean monthly water temperature, recorded over a 17-month period with a thermograph, usually varies between 5 and 17°C, and is very closely correlated with the mean monthly air temperature recorded at Valley, Anglesey; this can therefore be used to provide a continuous record. Long periods of hard frost are rare in this area, and it is because of this that we have been able to do so much culture work between tide marks.

TREATMENT OF TILES FROM BREEDING TANKS

The original purpose of the ground was to hold the tiles originating from the breeding tanks (see Chapter 2) until the spat were large enough to be removed from them, and then grown on in trays until they could be planted on the bottom. The general principles are similar to those used in oyster culture on the coast of France.

The tiles carrying the spat had to be cleared from Conwy in August or September. This was because, first, most of the tank space was required for mussel purification and, secondly, the growth and survival of spat kept at Conwy during the winter was poor. We found that a batch of 2000 could be transferred in a day. First thing in the morning the strings holding the bundles together were removed and replaced with galvanized wire. In later years a plastic cord was used when the bundles were originally assembled for dipping and, since this was not rotted by the sea water, it did not require replacing with wire. When ready, the bundles were loaded on a lorry, well packed with seaweed to prevent excessive drying, and transported to Tal-y-foel. Here, because it took some time to move this number down the beach, they were taken down in stages as the tide ebbed. As each batch was placed in shallow water, this reduced the risk of dessication. Before a firm pathway was built, yokes were found to be a relatively restful way of carrying the bundles (Plate 18).

The tiles were placed on slatted racks laid on crossbars between 7.5cm oak posts about 14cm above the ground. Each rack, measuring 90cm by 160cm overall, was made of two runners of 7.5 × 5.0cm or 10 × 3.7cm softwood timber, with 5 × 2.5cm slats fixed at 10cm centres. The racks were put up so as to make a continuous platform along low-water mark. Adjacent racks rested on a common crossbar. It was not necessary to fix the racks at all but a rail had to be provided on the inshore edge to avoid movement during the winter. The bundles of tiles were arranged in rows of three (Plate 19).

During the winter occasional visits were made to ensure that none of the structures had broken and to remove any drifting weed that might accumulate. In stormy weather a considerable layer of silt might be deposited on the tiles, but a wind from another quarter would usually remove this quite quickly. To ensure maximum survival it might be better to swill this off soon after it appears. Frost is not very common or intensive on that part of the Anglesey coast, but if a hard frost occurred at the time of low-water springs much of the lime-mortar would become detached from the top tiles.

The tiles were usually planted too late in the year for other sedentary organisms to grow to a competitive size by the winter, although sometimes Ascidians settled in considerable numbers in Conwy. As soon as the spring arrived, barnacles, tube worms, Ascidians and green seaweeds settled, grew, and started to smother the oyster spat. Generally, however, it was possible to start stripping the tiles before the situation became too serious; by June to August the spat were fingernail size, and sufficiently robust to repair the shell damage cause by stripping (Plate 20).

For stripping, the tiles were brought up to a hardened area at mid-tide level at low tide in the evening. A supply of tiles was collected from there early the next morning, to provide a stock for working on until the tide receded again; in this way it was possible to have tiles available for working for about 10 hours. Of course, a number of these manoeuvres had to be adopted because the breeding tanks were so far removed from the oysterage. In a commercial establishment designed for oyster breeding, the breeding tanks and the oysterage would be adjacent.

To strip the oysters from the tiles without damage, the spat should come away still attached to the lime-mortar. The success of the process depended on separating the mortar from the tile, rather than the oyster from the mortar. Useful tools were kitchen knives cut down to a 7.5cm blade, or, more usually, putty knives similarly ground to shape. The tiles were conveniently held against a stop on a sloping board fixed to a rest at waist height, and the detached spat rolled down the board into a wire basket below. When sufficient spat were detached to fill a tray, this was floated in the water so that those which had been damaged did not dry out. The detached spat had to be kept in trays because, first, they were very small and light and would be readily washed about by waves and tidal currents and, secondly, some were damaged in the stripping process. With exposed gills and mantle they would fall easy prey to crabs if unprotected.

The trays were made of 7.5 × 5.0cm or 10 × 3.7cm timber, and were made to fit in the same posts as the racks. If the tiles bore a reasonable crop, a rack of tiles would yield a tray of spat — about 5000 oysters. The base of the tray was usually of 1.2cm mesh woven wire, galvanized after manufacture, and the top, which had to bear little weight, was of 1.2cm mesh chicken wire. After assembly the trays, like the racks, were dipped. Formerly we used a liquid tar, then for a period we used a solid tar which was melted in a tank large enough to take a whole tray, and now we use a proprietary compound which does not require heating. Good results are also being obtained with plastic meshes, which compare very favourably in price with the high-quality woven wire it is necessary to use to obtain the required life. For the larger sizes of oysters which we wish to keep isolated, galvanized chain-link mesh is cheap and durable. If it is necessary to keep oysters in trays for some time they should be regularly moved on to the largest mesh that will retain them, because in this way the accumulation of mud, weed and dead shells in the trays is reduced. Economies in the number of posts can be obtained by stacking two or three trays; as the sides are solid it is best to separate them by a 5cm gap to allow a good circulation of water.

The fouling organisms which appeared during the spring after planting smothered some spat and reduced the growth and survival of others by impeding the free flow of water. Some of this could be avoided by stripping the spat as early as possible, since once the spat are in cages few fouling organisms can become established. The advantage of early removal from the collectors is shown by some data obtained in 1961. In April of that year a group of collectors carried on average 217 spat per bundle. About 15 bundles were stripped on each of three dates and the following table shows the yield of oysters in the following spring:

Date spat removed in 1961	Yield of oysters, March 1962	
	Number per bundle	Mean weight (g)
14 April	108	9.0
1 May	88	9.8
30 June	54	3.6

Another method of controlling competitors is to dip the bundles in a toxic solution. This method depends on the fact that oyster spat can remain more tightly closed than many other organisms. In the first trials tiles were dipped in a solution of 1 in 7000 of mercuric chloride for 15 minutes, and then stood in the dry for an hour before replacing in the water. This killed all the Ascidians, barnacles and Bryozoa and most of the tube worms, and left the oysters in excellent condition. However, mercuric chloride is a dangerous chemical and further trials were made with copper sulphate, commercial grades of which are quite cheap. A series of tests found that excellent results could be obtained by dipping the tiles into a solution of 4 parts of hydrated copper sulphate per 1000 parts of water, and then allowing to stand in the dry for 1-2 hours; this procedure can be easily carried out during the low-water period. The same solution is also used to remove the growth of green and brown weeds which occur on the wire meshes of the trays; dampening the weed, when first uncovered by the tide, with the copper sulphate solution, applied from a watering can, kills all of it.

MORTALITY OF OYSTERS AT TAL-Y-FOEL
In 1948 and 1951 a comparison was made of the death rate of young spat left at Conwy with others transferred to Tal-y-foel. The spat, which had settled on unglazed saucers, had a mean age of 14 days when they were transferred in August. The results showed that the number of deaths was very high — about 4 per cent per day — in the two following months, and on balance the difference in survival between those in the open sea and those in the comparative shelter of the Conwy tanks was negligible.

The survival on tiles which had been planted in the normal manner was estimated in 1945-46 and each of the years 1949-50 to 1953-54. The number of live spat was counted on several bundles of marked tiles at about the time of planting and the bundles were then mixed with the rest. At intervals the marked bundles were taken apart, and the spat counted.

In the whole period from planting to stripping mortality varied between 66 and 92 per cent. The initial high rate of 4 per cent per day declined rapidly to a low level by November and remained at this until March (Figure 34); after this, it

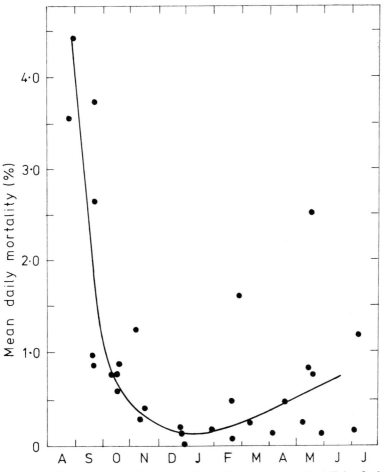

Fig. 34 The mean daily mortality of oysters at the Tal-y-foel oysterage during the year after metamorphosis. (From Walne, 1961.)

started to rise again but did not, in general, reach such high levels as in the first 2 or 3 months after settlement. The period of low mortality coincides with the period when the temperature is low and little or no growth is occurring, but whether the reduced mortality is a reflection of the reduced activity of the oyster or of the inactivity of predators and parasites is unknown.

In 4 out of the 6 years mortality was greater on the lower surfaces of the tiles than on the upper; this was probably due to the much greater growth of competitors on the underside. Some evidence was found that the oyster spat competed with each other, although the effect was not very great; for example, in 1953-54 tiles with 50-100 spat had a mortality of 88.2 per cent while those with 101-250 spat lost 92.7 per cent. Although this small difference was statistically significant, the increased productivity of the collector would make this small increase in mortality acceptable.

These losses are clearly very substantial and any reduction would have a considerable effect on yield. The common shore crab, *Carcinus maenas*, appears to be the only large potential predator but so far our investigations do not suggest that it is important. In September 1949, nine bundles of tiles carrying the spat of that summer were placed in a cage of 1.2cm mesh chicken wire similar tiles were planted in an unprotected rack alongside. The number of spat were counted at the time of planting and in the following August. By this time a few holes had appeared in the netting, so that the spat which had been protected when small were unprotected for the latter part of the period. The mortality rate in the protected spat (74.8 per cent) was so similar to that found in the controls (73.2 per cent) that it is unlikely that large predators were an important source of loss.

The losses amongst one-year-old oysters in the trays can be quite high. This has been mainly studied in hatchery-reared spat which have been stripped from the collectors in Conwy and allowed to heal before being transferred to Tal-y-foel. Of the very small spat, measuring 1-2cm, which had been stripped for about a month before transfer, 60 per cent died within 3 months, but with larger spat, measuring 3-4cm, the mortality was only about 35 per cent in the first 3 months. Thereafter the mortality rate dropped very considerably and in tray oysters over one year old it was rarely more than 10 per cent per year, and frequently only half this amount.

It is clear that mortality among oysters 1-3cm in size can often be severe, even in stocks which are protected from the larger predators and are free from excessive risk of covering with mud. Obviously the methods employed for them at present are not as satisfactory as those used for the younger stages, and further investigations should considerably enhance the productivity of artificial culture. Oysters of this size in trays are readily moved about by small waves and tend to accumulate on the inshore side, where some smothering must occur. It is likely that a greater degree of protection would be beneficial, if an adequate supply of water can be provided. It is sometimes suggested that hatchery-reared spat may not be sufficiently robust to withstand the rigours of life in the sea. At present it is difficult to refute this point, since there is little information available on the mortality rate of natural spat of this size in the sea; all we can say is that we have seen no evidence of it, and there is no indication of a sudden change in the death rate when spat are transferred from Conwy to the sea.

GROWTH OF OYSTERS AT TAL-Y-FOEL

A considerable number of observations have been made on the growth of oysters confined in cages (Plates 21, 22 and 23). This method has the advantage that the performance of a particular batch can be followed without having to rely on a sampling procedure. We are also fortunate in frequently being able to use animals of a known age-group. To reduce the natural variation in weight still further it has been our practice to select for study a group of individuals whose individual weight fell within a narrow band; this was sometimes as large as 5g, but more usually only 2-3g. The initial population was therefore composed of individuals of the same age with a weight which did not differ by more than about ±2g from the mean.

The greatest difficulty in studying growth in oysters is that there is no way of measuring the meat without destroying the oysters. Although the oysters could be readily weighed or measured at the oysterage, such estimates were heavily influenced by the shell which represents a large part of the total weight, and although suitable for gauging the total growth in a season were much less so for observing month to month variations. To detect the latter we divided a group of selected individuals into samples of about 50 each. Each sample was then confined in a separate compartment of a cage (we used standard cages fitted with nine compartments), and every 2 or 4 weeks a complete sample was brought back to Conwy for analysis. Such samples could be regarded as sub-samples of the main population and from them a picture could be built up of the seasonal growth.

Figure 35 illustrates the increase in the dry meat weight during five growing seasons, expressed as a percentage for successive 30-day periods while Figure 36 shows the % increase in nitrogen in 1967. These diagrams show not only the limits of the growing season but also the very considerable differences between seasons. The meat generally started to increase at the end of April or early May, and in the succeeding 1 or 2 months the dry meat could double in weight; thereafter growth might continue, but in some years there was little or no increase after mid-summer. Shell growth

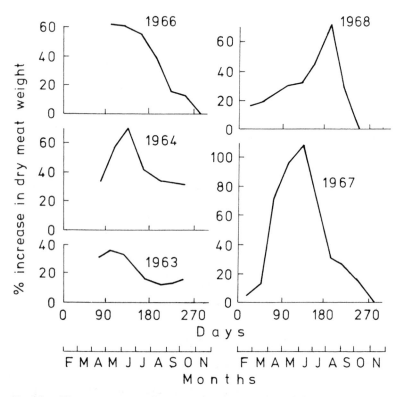

Fig 35   The percentage increase in dry meat weight in successive 30-day periods for *Ostrea edulis* at Tal-y-foel in 5 years.

172

follows a similar pattern, except that it normally starts several weeks later than the increase in meat.

The relation between the shell and the meat content of an oyster is of importance since it is the simplest way of judging the condition of the animal. A 'fat' oyster contains a larger

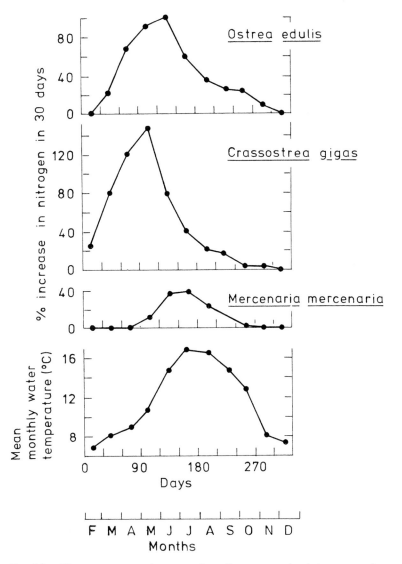

Fig. 36 The percentage increase in nitrogen content in successive 30-day periods for two species of oysters and for quahogs at Tal-y-foel in 1967. Nitrogen is indicative of the protein content of the animals.

proportion of meat than a thin one and this is obvious when samples of oysters are opened; a really good oyster fills a large proportion of the space between the two shells, whereas a poor one has plenty of spare space. This appearance can be given a practical expression by estimating the percentage of the internal space that is occupied by flesh. This is determined by taking a sample of oysters and, after removing all encrusting growth and mud, placing them in a displacement vessel and recording their volume. After the meats have been removed, the procedure is repeated with the shells and the difference between the two measurements gives the internal volume of the shell. The practice adopted with the meats has varied. At first, the volume of the meat, after allowing to drain in a standard manner, was measured by displacement. The difficulty with this is to standardize the method of draining the meat so that several operators can obtain the same result. In addition, the meats may contain different amounts of water, so from the biological aspects, the weight of dry meat has more meaning. For these reasons the formula which we now use to express the condition index is:

$$\frac{\text{Average dry weight of meat (g)}}{\text{Average volume between shells (ml)}} \times 1000.$$

The examination of many samples has shown that an average sample will have an index of about 90-100, whilst very well fattened oysters have an index of 120 or more. An index of 70-80 represents very poor condition.

The condition index can vary considerably from time to time and place to place, and Figure 37 shows its change in 5 years at Tal-y-foel. In general, condition has been found to be much better in the summer than in the winter. As well as observing the changing amounts of meat, we have examined alterations in its composition. One of the most regular features that has been seen is the seasonal change in ash content; this has regularly alternated between high values in winter and low values in summer, and is apparently closely linked to growth. The other major components have not displayed this regular change. Glycogen, an important food reserve, varies widely between about 13 and 42 per cent of the dry weight, but without any regularity. Nitrogen, which

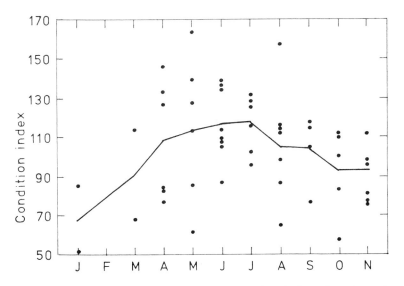

Fig. 37   The seasonal variation in the condition index of oysters at Tal-y-foel. Samples collected in the years 1963-68 with the exception of 1965.

indicates the protein content, is relatively constant between 6.1 and 8.4 per cent. The protein content expressed as a percentage of the dry flesh is particularly affected by the amount of the other constituents; in quantity it may increase or remain constant, though rarely does it decrease, but both ash and glycogen (which between them account for about half the dry flesh) may increase or decrease.

Very considerable differences in the annual growth have been recorded between different years. A long series of comparable data has been obtained for the years 1955-66 by sorting out by weight groups of 1½-year-old Conwy-reared oysters each winter, and setting them out in the standard trays. They have then been re-weighed the following winter. The average weight increment is illustrated diagrammatically in Figure 38 which demonstrates how, in a good year, the increase may be three times as much as in a poor year. The causes of this variation are at present unknown; in the opinion of the writer variations in food supply, perhaps in quality rather than quantity, are probably more important than climatic factors, although these may well act indirectly through the food. In favourable circumstances growth can be

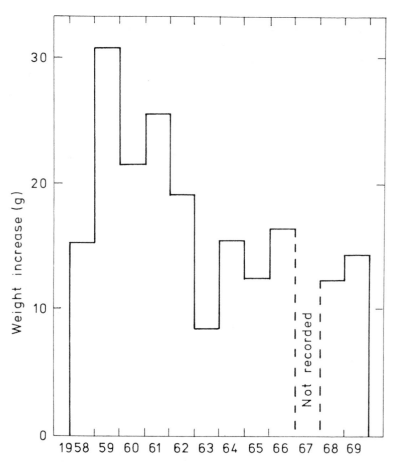

Fig. 38   The weight increment of 1½-year-old oysters at Tal-y-foel in successive years.

very rapid; for example, in late May and early June 1967 small oysters doubled in weight in about 30 days, although the water temperature was only about 12°C.

A further aspect which has received attention has been the effect of increasing tidal exposure. Six identical trays were placed individually 1 foot above the level of the mud at positions on the beach corresponding to different levels of exposure in the range 0-23 per cent. Each tray was stocked in April with 100 one-year-old and 100 two-year-old Conwy-bred oysters. The oysters were weighed again in October, and a straight line relation was found between the average weight increment and the degree of exposure. In both groups there was a weight increment of about 18g in those which were continually covered with water, and this declined steadily to about 4g in those which were exposed to the air for 23 per cent of the time. At the same time two further trays were erected on much longer posts so as to compare growth at the same tidal level, but at a different height above the beach; these gave the same results as the comparable trays close to the mud. These trials demonstrate that only a comparatively narrow band of the shore can be utilized without causing a marked reduction of the growth rate.

SELECTED REFERENCES
WALNE, P R, 1958. Growth of oysters (*Ostrea edulis* L.). *J. mar. biol. Ass. U.K.*, **37**, 591-602.
WALNE, P R, 1961. Observations on the mortality of *Ostrea edulis. J. mar. biol. Ass. U.K.*, **41**, 113-122.
WALNE, P R, 1970. The seasonal variation· of meat and glycogen content of seven populations of oysters *Ostrea edulis* L. and a review of the literature. *Fishery Invest., Lond.*, Ser. 2, **26 (3)**,35 pp.

Epilogue

It will be clear to the readers of this book that a great deal is now known about the requirements of bivalve larvae, and from this knowledge it is practical to define the techniques of large-scale hatchery culture. Further research can be expected to add refinements which are unlikely to add significantly to the costs of the process but would increase the yield and therefore bring about a reduction in the expense of producing a given number of oysters or clams. At the time of writing we are at an exciting stage in the development of hatchery culture. It is clear that it can be done, and many people believe that at some time in the future it will be an accepted part of the shellfish industry, but at present it remains to be demonstrated that it is yet profitable. The particular areas of uncertainty are the reliability of the method, month after month, year after year, the possible difficulties involved in carrying out the work at places other than Conwy, and the large-scale handling problems involved in the stage between the spat leaving the hatchery and becoming large enough to be relaid freely on the sea bed.

Economical solutions to many of the problems thrown up by increasing the scale of the operation may be expected to come from the prototype seed oyster unit built by the White Fish Authority at Conwy in 1964, and also from the increasing interest shown by commercial companies. Reference should be made at this point to the substantial development of shellfish hatcheries in the United States. These, based very largely on the American oyster, *Crassostrea virginica*, have mostly adopted a method which differs from the Conwy system in the provision of the food for the larvae. Little detailed information has been published but the basic principles are outlined in the US patent number 3 196 833. In these hatcheries, which are situated mainly on Long Island and in adjacent states, reliance is placed on the natural food in the sea water rather than on cultured food. The sea water is pumped into the hatchery where it first passes through a centrifuge which removes the particles that are too large to be eaten by the larvae, and also competing organisms. The clarified water is then held in shallow tanks for a period of 24 hours in a 'solarium'; this is a room with a roof of transparent plastic, and it is believed that there is a considerable multiplication of food organisms during the time the water is kept in bright sunlight. It has also been suggested that some purification of the water is effected by the ultra-violet rays of the sunlight. After this holding stage the water is run through into the larval culture vessels, which are treated in a similar manner to the methods described in this book. There is no doubt that this method works, and some substantial sized hatcheries, utilizing 50 or more larval rearing vessels, have been operating for 5 to 10 years, although data on their output and efficiency have not been published. There is a problem in that this type of operation depends ultimately on the natural food in the sea water which may not be adequate in either quality or quantity all the time, and in some places it may always be inadequate. In the area where this technique was developed (Great South Bay on Long Island) the water is exceptionally rich in small flagellates, and in the writer's experience probably richer than anywhere in the British Isles. If it can be utilized in some favoured places it is clearly a cheap method of providing enriched water, and could be useful in feeding spat, but it is probably unwise to rely solely on this source of food for larvae without a very considerable knowledge of the flagellate population of the chosen area.

Once hatcheries can be shown to be commercially profitable, it is clear that, with continual advance in knowledge, their potential for producing very large quantities of seed stock is considerable. What is the likely size of the market? Estimates of this lie in the field of commercial expertise, although some hints can be gleaned from considering the size of the industry in the past in this country and at present elsewhere. According to the MAFF Fishery Statistics the landings of oysters in 1920 reached 39 millions; this gradually declined, reaching 6 or 7 million in the late 1940s and about 5 million in 1967. This decline is associated with difficulties in production and not with any inability of the industry to sell its product. Although there are formidable problems to be overcome (for example, disease, pests, and sewage and industrial pollution) these can all be faced if the industry can be assured of an adequate and regular supply of seed stock.

The production and consumption of bivalves in Great

Britain is only a fraction of that found in Western Europe and the United States, if the landings are compared with the human population (Table 8). Although these figures do not strictly represent consumption because they will be distorted by international trade, they are, nevertheless, a guide to the interest found in each country in this type of crop.

An indication of the potential for international trade was given in the issue of *Cultures Marines* for January 1969, which showed that the importation of oysters into France for the 1967-68 season was as follows:

	Tonnes
Flat oysters, less than 40g	1182
Flat oysters, more than 40g	143
Portuguese oysters, less than 35g	6908
Portuguese oysters, more than 35g	224
Total	8457

If the low value of £100 per tonne is given to these oysters, this trade alone had an annual value of nearly one million pounds and helped to satisfy the demand in a country where some thousand million oysters are eaten each year. The annual value of the bivalve crop in the United States is very considerable: 7-10 million dollars for quahogs and 30-40 million dollars for oysters.

The countries mentioned in the last few paragraphs all have a standard of living similar to or higher than that in Britain, and their way of life is not so unfamiliar as to suggest that shellfish, either in the shell or processed, could not find a good market in this country. The need is for adequate stocks, with an assurance of annual replenishment, on which active marketing can be based. It is the writer's hope that this book will help to spread information on one way in which this might come about.

TABLE 1

The filtration rates, at 20-21°C and at a flow rate of 200ml per minute, of five species of bivalves in successive 1cm size groups. 'Length' in *Ostrea* refers to average diameter, in *Mercenaria* and *Venerupis* to the antero-posterior axis, and in *Crassostrea* and *Mytilus* to the measurement from the umbo to the opposite edge of the valve.

Length (cm)	Filtration rate per animal (ml per minute)				
	Ostrea edulis	Crassostrea gigas	Venerupis decussata	Mercenaria mercenaria	Mytilus edulis
3	—	—	46	41	—
4	69	96	53	50	41
5	86	105	62	59	53
6	103	114	—	67	64
7	122	120	—	73	76
8	138	127	—	—	—
9	—	133	—	—	—
10	—	141	—	—	—

TABLE 2

Stock enrichment solutions for the culture of *Isochrysis galbana* and other algae.

1. $FeCl_3.6H_2O$ 2.60g
    $MnCl_2.4H_2O$ 0.72g
    H_3BO_3 .... 67.20g
    E.D.T.A. (Na salt) 90.00g
    $NaH_2PO_4.2H_2O$ 40.00g
    $NaNO_3$ 200.00g
    Trace metal solution 2.0ml
    Distilled water .... to 2 litres
    1ml is added to each litre of sea water

2. The trace metal solution has the following composition:

    $ZnCl_2$ 2.1g
    $CoCl_2.6H_2O$ 2.0g
    $(NH_4)_6Mo_7O_{24}.4H_2O$ 0.9g
    $CuSO_4.5H_2O$ 2.0g
    Distilled water .... to 100ml
    It is necessary to acidify this solution with HCl to obtain a clear liquid

3. Vitamin: stock solution

    B_{12} 10mg
    B_1 (Thiamine) 200mg
    Distilled water .... to 200ml
    10ml are added to each 100 litres of sea water.

TABLE 3

Formula for artificial sea water (from Lyman and Fleming, 1940).

NaCl 23.476g
$MgCl_2$ 4.981g
Na_2SO_4 .... 3.917g
$CaCl_2$ 1.102g
KCl 0.664g
$NaHCO_3$ 0.192g
KBr 0.096g
H_3BO_3 .... 0.026g
$SrCl_2$ 0.024g
NaF 0.003g
Water to 1kg

TABLE 4

The chemical composition (μg per 10^6 cells) of four species of algae. All were grown in 20 litre cultures; those marked * were in semi-continuous culture at Conwy and bacteria were present in the cultures.

Cell density index = $\dfrac{\text{dry weight}}{\text{volume of packed fresh cells}}$

Species	Age of culture in days	Nitrogen	Carbohydrate	Triglyceride	Cell density index
Dunaliella tertiolecta	12	1.1	35.1	1.5	27
	19	1.7	61.2	4.1	32
	33	1.4	102.4	16.0	38
Phaeodactylum tricornutum	9	0.4	2.0	0.6	11
	17	0.4	5.1	6.0	19
	6*	1.4	2.7	—	19
Monochrysis lutherii	13	0.7	9.4	2.9	40
	21	0.6	10.0	3.9	40
Isochrysis galbana	7*	1.2	2.3	—	16
	5*	1.2	2.5	—	9
	56*	1.3	3.2	—	17
	48*	1.5	3.7	—	17
	20*	1.4	2.7	—	23
	17	0.6	1.3	0.6	14
	25	0.6	1.4	1.4	12

TABLE 5

Index of food assimilation, related to temperature. Calculated from experiments in which *Isochrysis* labelled with ^{32}P was fed to oyster larvae.

Cells of Isochrysis per µl	Temperature (°C)				
	17	20	23	25	26
10	19	27	35	40	43
25	38	55	70	82	87
50	56	80	108	120	126
100	70	100	128	150	158

TABLE 6

The growth of *Ostrea edulis* spat when fed on various foods compared with controls grown on *Isochrysis*. The index has been calculated by dividing, for each experimental series, the largest mean size in the series at the measurement made on the nearest to 21 days from the commencement of the test, by the mean size of the control on that day.

Species	Index of food value	
	Individual experiments	Average
Monochrysis lutheri	1.70, 1.03	1.36
Chaetoceros calcitrans	1.28	1.28
Tetraselmis suecica	1.42, 1.06, 1.12	1.20
Skeletonema costatum	0.93, 1.09	1.01
Isochrysis galbana	—	1.00
Dicrateria inornata	0.85, 1.04	0.94
Cryptomonas sp.	0.54, 0.74	0.64
Cricosphaera carterae	0.41, 0.83, 0.61	0.62
Chlorella stigmatophora	0.65, 0.56	0.60
Phaeodactylum tricornutum	0.62, 0.43, 0.73	0.59
Olisthodiscus sp.	0.47, 0.71	0.56
Nannochloris atomus	0.60, 0.60, 0.41	0.54
Chlorella autotrophica	0.37, 0.66	0.52
Pavlova gyrans	0.69, 0.32	0.50
Micromonas minutus	0.50, 0.40, 0.42	0.44
Dunaliella euchlora	0.40	0.40
Dunaliella tertiolecta	0.36, 0.42	0.39
Chlamydomonas coccoides	0.26, 0.33	0.30

TABLE 7

The mean initial and final weight (in grammes) of the three bivalve species tested in different parts of the country.

Area	Species		
	Pacific oyster	*American hard clam*	*Chilean mussel*
Loch Tournaig	15.9- 45.9	0.86-1.9	0.53- 3.3
Linne Mhuirich	15.6- 44.9	0.88-2.5	0.51- 6.5
Menai Bridge	16.2- 54.0	0.80-2.0	0.46- 3.7
Tal-y-foel	15.1- 65.9	0.84-3.8	0.47- 7.4
Holyhead	15.7- 76.8	0.68-1.8	0.46- 4.3
River Yealm	15.7- 74.7	0.89-3.7	0.51-11.2
Newtown River	16.3-104.2	-6.0	0.48- 8.1
River Crouch	15.0- 80.5	0.92-4.8	0.48- 7.1
Althorne Creek	15.0- 77.3	0.87-3.9	0.47- 6.5
River Roach	14.8- 77.4	0.91-5.0	0.47- 6.7
Average	15.5- 70.2	0.85-3.54	0.48- 6.48

TABLE 8

The average annual landings, by weight, of bivalves, 1960-63 (from FAO statistics) related to the human population and the length of coastline.

Country	Landings (thousands of metric tonnes)	Population (millions)	Coastline (miles)	Landings (thousands of metric tonnes)	
				Per million population	Per 1000 miles
USA	567	193	12 383	2.9	45.8
France	106	48	1 750	2.2	60.6
Holland	81	12	1 076	6.8	75.3
UK	3.9	54	4 910	0.07	0.80

Subject Index

187

Other books published by Fishing News Books Limited
Farnham, Surrey, England

Free catalogue available on request
A living from lobsters
Advances in aquaculture
Aquaculture practices in Taiwan
Better angling with simple science
British freshwater fishes
Coastal aquaculture in the Indo-Pacific region
Commercial fishing methods
Control of fish quality
Eel capture, culture, processing and marketing
Eel culture
European inland water fish: a multilingual catalogue
FAO catalogue of fishing gear designs
FAO catalogue of small scale fishing gear
FAO investigates ferro-cement fishing craft
Farming the edge of the sea
Fish and shellfish farming in coastal waters
Fish catching methods of the world
Fish farming international No 2
Fish inspection and quality control
Fisheries of Australia
Fisheries oceanography
Fishery products
Fishing boats and their equipment
Fishing boats of the world 1
Fishing boats of the world 2
Fishing boats of the world 3
Fishing ports and markets
Fishing with electricity
Fishing with light
Freezing and irradiation of fish
Handbook of trout and salmon diseases
Handy medical guide for seafarers
How to make and set nets
Inshore fishing: its skills, risks, rewards
International regulation of marine fisheries: a study of regional
    fisheries organizations
Marine pollution and sea life
Mechanization of small fishing craft

Mending of fishing nets
Modern deep sea trawling gear
Modern fishing gear of the world 1
Modern fishing gear of the world 2
Modern fishing gear of the world 3
Modern inshore fishing gear
More Scottish fishing craft and their work
Multilingual dictionary of fish and fish products
Navigation primer for fishermen
Netting materials for fishing gear
Pair trawling and pair seining—the technology of two boat fishing
Pelagic and semi-pelagic trawling gear
Planning of aquaculture development—an introductory guide
Power transmission and automation for ships and submersibles
Refrigeration on fishing vessels
Salmon and trout farming in Norway
Salmon fisheries of Scotland
Seafood fishing for amateur and professional
Ships' gear 66
Sonar in fisheries: a forward look
Stability and trim of fishing vessels
Testing the freshness of frozen fish
Textbook of fish culture; breeding and cultivation of fish
The edible crab and its fishery in British waters
The fertile sea
The fish resources of the ocean
The fishing cadet's handbook
The lemon sole
The marketing of shellfish
The seine net: its origin, evolution and use
The stern trawler
The stocks of whales
Training fishermen at sea
Trawlermen's handbook
Tuna: distribution and migration
Underwater observation using sonar